T0212494

Lecture Notes in Computer Science   10026

Commenced Publication in 1973
Founding and Former Series Editors:
Gerhard Goos, Juris Hartmanis, and Jan van Leeuwen

More information about this series at http://www.springer.com/series/7411

Selma Boumerdassi · Éric Renault
Samia Bouzefrane (Eds.)

Mobile, Secure, and Programmable Networking

Second International Conference, MSPN 2016
Paris, France, June 1–3, 2016
Revised Selected Papers

 Springer

Editors
Selma Boumerdassi
CNAM/CEDRIC
Paris
France

Samia Bouzefrane
CNAM/CEDRIC
Paris
France

Éric Renault
Institut Mines-Télécom – Télécom SudParis
Evry
France

ISSN 0302-9743   ISSN 1611-3349   (electronic)
Lecture Notes in Computer Science
ISBN 978-3-319-50462-9   ISBN 978-3-319-50463-6   (eBook)
DOI 10.1007/978-3-319-50463-6

Library of Congress Control Number: 2015952766

LNCS Sublibrary: SL5 – Computer Communication Networks and Telecommunications

This Springer imprint is published by Springer Nature
The registered company is Springer International Publishing AG
The registered company address is: Gewerbestrasse 11, 6330 Cham, Switzerland

Preface

The rapid deployment of new infrastructures based on network virtualization and cloud computing triggers new applications and services that in turn generate new constraints such as security and/or mobility. The International Conference on Mobile, Secure and Programmable Networking (MSPN) is aimed at providing a top forum for researchers and practitioners to present and discuss new trends in networking infrastructures, security, services, and applications while focusing on virtualization and cloud computing for networks, network programming, software-defined networks (SDN) and their security. In 2016, MSPN was hosted by CNAM Paris, which is one of the oldest teaching centers in Paris.

The call for papers resulted in a total of 37 submissions from all around the world. Every submission was assigned to at least three members of the Program Committee for review. The Program Committee decided to accept 17 papers. The accepted papers are from: Algeria, China, France, Greece, India, Ireland, Italy, Morocco, Tunisia, and Vietnam. One intriguing keynote from Prof. Ruben Milocco of the University of Comahue, Argentina, completed the technical program.

We would like to thank all who contributed to the success of this conference, in particular the members of the Program Committee (and the additional reviewers) for carefully reviewing the contributions and selecting a high-quality program. Our special thanks go to the members of the Organizing Committee for their great help.

We hope that all participants enjoyed this successful conference, made a lot of new contacts, engaged in fruitful discussions, and had a pleasant stay in Paris, France.

June 2016

Selma Boumerdassi
Éric Renault
Samia Bouzefrane

Organization

MSPN 2016 was organized by the CEDRIC laboratory of CNAM Paris and the Wireless Networks and Multimedia Services Department of Télécom SudParis (a member of Institut Mines-Télécom) in cooperation with IFIP Working Group 11.2 on Pervasive Systems Security.

General Chairs

Selma Boumerdassi   CNAM, France
Éric Renault   Institut Mines-Télécom – Télécom SudParis, France
Samia Bouzefrane   CNAM, France

Publicity Chair

Filippo Gaudenzi   Università degli Studi di Milano, Italy

Organizing Committee

Linda Chamek   CNAM, France
Thinh Le Vinh   CNAM, France
Madina Omar Eleyeh   CNAM, France
Youcef Ould Yahia   CNAM, France

Technical Program Committee

Nadjib Achir   University of Paris 13, France
Rachida Aoudjit   University Mouloud Mammeri of Tizi-Ouzou, Algeria
Hanifa Boucheneb   Ecole polytechnique de Montréal, Canada
Emmanuel Conchon   University of Limoges, France
Yuhui Deng   Jinan University, China
José María de Fuentes   Carlos III University of Madrid, Spain
Makhlouf Hadji   IRT SystemX, France
Viet Hai Ha   University of Hue, Vietnam
Li Li   Wuhan University, China
Malika Belkadi   University of Tizi-Ouzou, Algeria
Sjouke Mauw   University of Luxembourg, Luxembourg
Alessio Merlo   University of Genoa, Italy
Pascale Minet   Inria, France
Hassnaa Moustafa   Intel, USA
Paul Mühlethaler   Inria, France
Thao Nguyen   NIST, USA
Abdelkader Outtagarts   Nokia Bell Labs, Villarceaux, France
Jihene Rezgui   LRIMA Lab, Maisonneuve, Canada

Leila Saidane	Université de Manouba, Tunisia
Xuan-Tu Tran	Vietnam National University, Vietnam
Stefano Zanero	Politecnico di Milano, Italy
Weishan Zhang	University of Petroleum, China

Additional Reviewers

Mohamed Bouatit	CNAM, France
Linda Chamek	CNAM, France
Aravinthan Gopalasingham	Nokia Lucent Bell Labs Villarceaux, France
Sonia Ikken	Institut Mines-Télécom – Télécom SudParis, France
Van Khang Nguyen	Institut Mines-Télécom – Télécom SudParis, France
Van Long Tran	Institut Mines-Télécom – Télécom SudParis, France
Youcef Ould Yahia	CNAM, France

Sponsoring Institutions

Conservatoire National des Arts et Métiers, Paris, France
Institut Mines-Télécom – Télécom SudParis, Évry, France

Contents

Efficient Intermediate Data Placement in Federated Cloud Data Centers Storage

Sonia Ikken[1,2(✉)], Eric Renault[1], Amine Barkat[2,3], M. Tahar Kechadi[4], and Abdelkamel Tari[2]

[1] Telecom SudParis, Évry, France
{sonia.ikken,eric.renault}@telecom-sudparis.eu
[2] Faculty of Exact Sciences, University of Bejaia, 06000 Bejaïa, Algeria
tarikamal59@gmail.com
[3] Politecnico di Milano, Milano, Italy
amine.barkat@polimi.it
[4] UCD School of Computer Science and Informatics, Belfield, Ireland
tahar.kechadi@ucd.ie

Abstract. The goal of cloud federation strategies is to define a mechanism for resources sharing among federation collaborators. Those mechanisms must be fair to guaranty the common benefits of all the federation members. This paper focuses on intermediate data allocation cost in federated cloud storage. Through a federation mechanism, we propose a mixed integer linear programming model (MILP) to assist multiple data centers hosting intermediate data generated from a scientific community. Under the constraints of the problem, an exact algorithm is proposed to minimize intermediate data allocation cost over the federated data centers storage, taking into account scientific users requirements, intermediate data dependency and data size. Experimental results show the cost-efficiency and scalability of the proposed federated cloud storage model.

Keywords: Big data workflow system · Intermediate data · Storage federation · MILP

1 Introduction

1.1 Challenge

In recent years, the advances in cloud computing technologies has given birth to big data workflow system (BDWS). The use of cloud infrastructure for those platform[1] facilities the composition of the individual operation to provide essential support to data analytics, high performance computing and on-line data storage. Therefore, all these operations are easier to deploy, manage and execute over

---

[1] Example of this platforms: Hadoop-MapReduce [1], NoSQL [2], OpenStack-NovaOrchestration [3] etc.

© Springer International Publishing AG 2016
S. Boumerdassi et al. (Eds.): MSPN 2016, LNCS 10026, pp. 1–15, 2016.
DOI: 10.1007/978-3-319-50463-6_1

cloud which offer several opportunities such as virtualization, efficient resource utilization and aggregation power. BDWS processing on clouds may involve hundreds of cloud storage servers accessing intermediate data and results, and this leads to the generation of a massive amount of intermediate data sets requests to be hosted and managed. BDWS are very complex, one operation might require many intermediate data sets for processing. Therefore, some intermediate data sets are always used together by many operations, this implies the existence of dependency between each other. Handling large intermediate data dependency in cloud storage is important for such operations and need a long time for execution since those intermediate data dependency needs to process data from different data centers. Some intermediate data sets are very large to be relocated efficiently, this operation must take into account dependencies between intermediate data sets in selecting their locality, therefore, moving intermediate data sets to the most efficient data center becomes a challenge. Furthermore, scientific users share important intermediate data dependencies for cooperation and reproduction of new intermediate data based on data provenance. Capturing provenance for recalculating intermediate results reveals data retrieval from cloud storage resources which their size continuously grows depending on intermediate data dependency size, hence, the total storage cost increases.

A common feature in cloud computing infrastructures is collocation of storage and computing nodes for more efficient data management. Although, it is oblivious to store each amount of intermediate data in such a cloud environment that offers abundant storage resources. The amount of data size generated, transferred and shared by BDWS in a single Cloud storage provider significantly increases along with colocation Cloud infrastructure (HDFS-Hadoop [4], Swift-OpenStack [5], AWS [6] etc.). BDWS places higher performance demands on the computational systems and storage requirements. Each cloud storage provider has his own solution and storage offers with variables prices depending on the users need. The adoption of cloud storage of single provider can unveil the unforeseen cost of intermediate data access and inefficacy in using storage and compute resources. In addition, some problems may arise, such as: unavailability of storage services due to some internal problems, risks of losing intermediate data, vendor lock-in problem and many others. Therefore, the problem of using a single cloud provider could be solved using BDWS and cloud storage federation for more cost-effectiveness for intermediate data allocation.

1.2   Context

A storage federation allows users going to take the simplistic migration, backup, synchronization, data and resources allocation across connected cloud storage systems. It collaborates to reduce the cost through outsourcing data storage according to a negotiated dynamic pricing plan [7]. However, the dynamic price mechanism used needs the cloud exchange the price and the transaction demand to the cloud coordinators, ceaselessly until an allocation-demand balance is achieved. During this process, the price needs to be updated during exchanges.

Federated cloud storage focuses on outsourcing part of the data load by borrowing storage resources from foreign cloud storage when the home cloud storage is overloaded or has limited capacity, also insourcing data requests by renting resources to other cloud storage when the home cloud storage is free. Outsourcing intermediate data storage can relieve pressure on storage resources capacity of a single provider while reducing costs of users. For scientific users, it is important to use federated cloud storage to converge to an optimal intermediate data allocation plan with a minimal cost.

1.3 Scop of the Work

The contribution of this paper is an optimization model that solves the problem of allocating distributed intermediate data in a federated cloud storage. Our model is based on Mixed Integer Linear Programming (MILP) optimization. In this solution, we proposed an exact algorithm based on MILP that takes into account dependencies type (valuable and unnecessary correlation) of this intermediate data for making decisions, and it reallocate intermediate data with dependencies in one data center to reduce total storage cost for scientific users demands. In our model, we propose to store all the generated intermediate data to avoid costs of processing for regeneration of some deleted intermediate data, since all intermediate data are generated at each workflow phases from local data center storage. Our goal is to decrease the total cost of storing data in the Cloud federated, taking advantage of different available offers from participated providers. In each time a set of intermediate data is generated, our algorithm could be used to provide and optimal storage plan that minimizes costs and simultaneously satisfies scientific users requests and intermediate data dependencies.

1.4 Paper Organization

The paper is organized as follows. Section 2 outlines the related work. Section 3 describes the model system based on the cloud storage federation scenario with target assumptions. Based on this system model, Sect. 4 derives exact optimization approach for allocating intermediate data on federated Cloud storage. Section 5 shows and discuss the results simulation obtained with set of example scenario. Finally, the Sect. 6 concludes this work and presents some future work.

2 Related Work

Data storage optimization has received a lot of attention in data workflow systems, and some previous works used the cost model to provide certain features from cloud storage scenario. The works presented in [8–10] are closer to our focusing problem. Authors in [8] elaborated a cost-effective strategy for storing intermediate data workflow in a single cloud storage provider using Amazon based fixed pricing. The proposed model focuses on running scientific workflow system in cloud and automatically deciding whether intermediate dataset should be stored or deleted in the cloud storage provider considering users tolerance of computation delays. Since our allocation strategy does not remove intermediate data sets before the end of workflow system processing, as shown in the introduction, all intermediate data sets are stored in multi-data centers involving a cloud storage federation with a cost storage optimization. This differs considerably from [8] which can not be compared to our model. The authors in [9] presented a matrix based k-means clustering strategy for data placement in scientific cloud workflow. The authors stress the movement of large volumes of data that can automatically allocated among data centers based on the data dependencies. The optimization is done only at data movement level and the authors was not defined a storage cost optimization during the intermediate data placement which differs from our approach. In addition, our approach takes into account the type of dependency in order to further optimize the data movement and storage cost. The authors in [10] is closer to the work in [9]. The authors presented a data placement strategy based on data dependency clustering for scientific workflow in heterogeneous cloud. The storage cost in this work is not considered and they focused only on the cost of data re-distribution. A cloud storage cost based selection decision are presented in [11–13]. The authors in [11] proposed an optimization problem of selecting the best storage services and take into consideration applications requirements and users priorities. A total storage cost is solved and workload requirements are not addressed in this work. In [12], the authors identified an adaptive cost optimization system for multi-cloud storage to decrease the storage cost, a compression and placement algorithm are used to reduce data cost in cloud storage, however, data correlation is not considered on the optimization cost model. The authors in [13] proposed a storage cost optimization based on linear programming model using multiple public cloud storage providers. This work differs from our model since the type of data and collaboration aspect based cloud providers are not envisaged.

3   System Model

The target of this work is to build a federated cloud storage model to allocate optimally intermediate data workflow. To derive the model, a multiple assumptions is simplified regarding to the storage federation scenario which is focused on pricing negotiation between cloud storage services which are already federated that we assumed to be cheaper than local prices without federation. This can be explained by a large intermediate data set access requests noted by IDi and their respective size noted by $size_i$ that will be allocated to achieve competitive storage services, maximum storage resources utilization and prevent intermediate data lock-in.

3.1   Scenario Assumptions

The scenario introduced in Fig. 1 illustrates the assumed federated data centers storage D geographically distributed which provides on-line mass storage

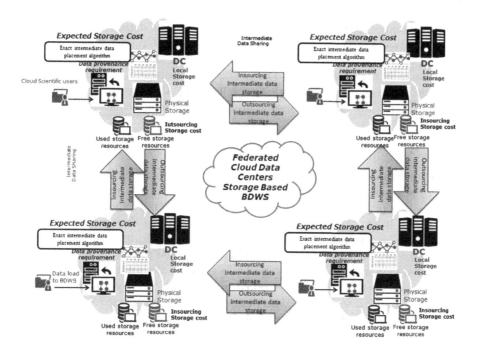

Fig. 1. Federated data centers storage scenario

to scientific users for the set of intermediate data IDi workflow placement. d, d' designate local data center storage where intermediate data are temporary stored. Indeed, the set of federated data center storage D are aggregated and interconnected using native peer-to-peer communication to shift intermediate data workloads from busy disks entities to those with available capacity, and efficiently use storage resources of the data center and balance intermediate data placement requirements. The intermediate data sets placement decision on one or multiple data centers of the storage federation involves the use of data centers converged in local and storage federation for processing and reutilization internally or by the other scientific users. After generating intermediate data set locally noted by IDN_d on each data center that combined in a source data center matrix noted by $DCsource_d$, a memory-buffer temporary stores these intermediate data sets before their placement decision.

Each federated data center storage k knows the storage space quotas SCF_k made available by the other federation collaborators as well as their internal capacity. The federated data center storage will use the expected storage cost and the proposed algorithm according to the data dependency to allocate their intermediate data resources to the federated data center storage. The storage selection decision has to lead to minimum data storage cost with intermediate data dependency constraint (one data center hosting intermediate data dependency sets) and maximum storage utilization for the federated data centers.

Each cloud storage provider which hosts federated data centers proposes cooperatively a storage cost (\$ per GB) noted by OSC_k, and a local storage cost for the intermediate data hosted locally noted by LSC_d. A storage resource federation price varies according to the idling space storage S_{idle} of the federated data centers, hence, we adapt a pricing mechanism [14] to our storage federation cost to dynamically set their insourcing prices for intermediate data storage space:

$$S = (SCmax - (S_{idle} + S_{free}))\backslash SCmax * (S_{price} - ME_{price}) + ME_{price} \quad (1)$$

Each federated data center provides free cloud storage quota noted by S_{free} to every scientific users. Therefore, the free quota price is reduced from the storage space pricing of the maximum capacity. The minimum effective price ME_{price} is a basic storage infrastructure resources pricing that providers do not fall below. The affected cost S_{price} for the end-scientific users is fixed which varies according to standard on-demand cloud storage pricing plan. Moreover, the maximum storage federation capacity is given by the totality of data center storage quota offered by each one, noted by $SCmax_k$. A very important point to consider during intermediate data placement on the storage federation is data transfer in or from other federation collaborators. It should be mentioned that in most cloud storage services the pricing of data transfer is more expensive than the data storage itself, so this is an important cost factor that must be considered in our optimization problem. Therefore, transfer cost of insourcing and outsourcing are noted respectively by ITC_k and OTC_k. The outsourcing and local storage cost OSC_k, LSC_d are updated using Eq. 1, that depend on their available outsourcing versus local capacity and the minimum effective storage price of each generated intermediate data set.

3.2   Intermediate Data Dependency Matrix

Each federated data center storage k receives intermediate data allocation requests. When new intermediate data set are generated by an instance workflow, the scientific users dynamically compute the dependencies of a new intermediate data sets which are hosted temporally on the local data centers. To represent the dependency for each intermediate data sets we define a binary integer matrix with symmetric value noted by DEP, for each coefficient of the matrix the value noted by Dep_{ij} is defined as:

$$Dep_{ij} = \begin{cases} 1 & \text{dependency between two intermediate data sets } i,j \\ O & \text{otherwise.} \end{cases} \quad (2)$$

Figure 2 illustrates an example of intermediate data dependencies and the corresponding matrix. Each intermediate data has a dependency with itself and some others. The scientific users are in charge of publish-subscribe operations of BDWS, therefore, the inputs of the optimization model are a matrix Dep_{ij} which is introduced by the scientific user identifying references between intermediate data request.

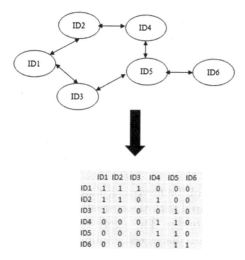

Fig. 2. Intermediate data dependency matrix

However, various kinds of dependencies can be specified between the intermediate data during a freak execution of a BDWS. Scientist user may encounter errors in the execution that will cause unnecessary dependencies that need to be adjusted or re-run, hence, some intermediate data dependencies are not valuable. In order to consider this situation, a float parameter λ_i^j is needed that denoting scientific users's tolerance of intermediate data dependencies of i and j:

$$\lambda_i^j = \begin{cases} O & \text{indicates scientific users have no tolerance to process } i,j \text{ independently} \\ 1 & \text{otherwise.} \end{cases} \quad (3)$$

The generated intermediate data sets that represent the dependencies or affinities among them with a positive value of λ_i^j operate over a set of I/O request sequencing between each other in the selected data centers. Those some operations can be involve remote access request: adjustment operation, re-processing or re-utilization operations. For each I/O access, there is a cost noted by $IOPCost_{i,j}$. The values of the dependency matrix DEP is dynamically maintained at each workflow phases. For each value of the coefficients i,j, the value of λ_i^j parameter is generated. The amount and the size of intermediate data dependencies will feed the expected storage cost when allocating intermediate data storage IDi. Our proposed optimization model collaborates by sharing their respective storage resources and dynamically adjust their hosting capacities according to their intermediate data allocation requirements.

4   Exact Intermediate Data Placement Algorithm

The proposed exact intermediate data placement algorithm is an MILP approach through the inclusion of valid conditions expressed in the form of constraints or inequalities. Finding the optimal intermediate data placement requires the computation of storage cost for each possible intermediate data placement solution from one workflow phase on a local and federated data centers storage until it reaches the number of phases for the total storage cost. The number of phase is given randomly before the start of the algorithm:

1. When a scientific user that processes intermediate data on local data center wants to send an allocation request to the storage federation, each data center storage from federation collaborators can expose their insourcing storage cost based on the pricing storage estimation noted by Eq. 1 with respect to the federation.
2. At each phase, the cost of the newly calculated intermediate data sets placement solutions is compared with the currently lowest cost placement in each federated data centers storage. The algorithm terminates after the set of all relevant intermediate data placement solutions have been checked. The algorithm is performed under intermediate data owner requirements and capacities constraints, so the choice of a set of data centers storage to be involved in hosting the intermediate data sets.
3. Furthermore, the proposed algorithm tries to keep these intermediate data dependencies in one data center while saving their storage and movement cost. By placing intermediate data with their dependencies, our algorithm attempts to minimize the data movement during the execution of workflow phases. This placement strategy can prevent intermediate data gathering from source data center to an outsourcing data center that relevant data are outsourced. The intermediate data sets with dependency tolerance will be optimized according to the cost of I/O-demands on federated data centers storage.
4. The proposed MILP cost model based exact placement algorithm uses pricing storage mechanism at each data center and initializes the input parameters (Table 1), and seeks all possible intermediate data placement solutions at each phase. The pricing storage computation, the dependency matrix and λ_i^j parameter are updated. At the end of the last phase, the computation of the total storage cost (Eq. 4) is performed.

The problem is solved on each workflow phases when a new intermediate data set is generated from a local data center. Such cost is assumed to change according to the federation feature.

In the following we introduce the used of the bivalent variables 0–1 integer x_{id}^k, tell us which data center storage d that generate intermediate data i to allocate it in federated data center k. A glossary of all the notations used and their descriptions in the proposed model is shown in Table 1. Using these notations, the global objective function is given by Eq. (4):

$$MinCost = \sum_{idk}^{d \neq k} x_{id}^k \cdot size_i \cdot (OSC_k + ITC_k + OTC_d)$$

$$+ \sum_{idk}^{d=k} x_{id}^k \cdot size_i \cdot LSC_k + \sum_{ijdd'k}^{d \neq k} y_{ijdd'}^{kk'} \cdot Dep_{ij} \cdot \lambda_i^j \cdot IOPC_{i,j} \qquad (4)$$

subject to several linear and integrity constraints expressed respectively by Eqs. (5) to (15):

For all intermediate data sets IDi generated and stored temporarily in each data center d there is one or more outsourcing data center storage k in the federation that will store the set of IDi:

$$\sum_{ik} x_{id}^k = IDN_d \qquad \forall d = 1, ..., D; \quad d \neq k \tag{5}$$

For each intermediate data set i generated and stored temporarily in data center d there is only one data center storage hosting i:

$$\sum_{dk} x_{id}^k = 1 \qquad \forall i = 1, ..., ID \tag{6}$$

For each generated intermediate data sets i,j with $Dep_{ij} = 1$ and with no dependency tolerance $\lambda_i^j = 0$, their storage allocation will be in the same federated data center storage k:

$$x_{id}^k + x_{jd'}^k = 2 \qquad \forall i, j = 1, ..., ID; \quad \forall k, d, d' = 1, ..., D \tag{7}$$

For each intermediate data dependency of i, j with a dependency tolerance $\lambda_i^j = 1$, which is specified by scientific users from local data center storage d, d', their storage allocation will be in the different federated data centers storage:

$$x_{id}^k + x_{jd'}^k \leq 1 \qquad \forall i, j = 1, ..., ID; \quad \forall k, d, d' = 1, ..., D \tag{8}$$

To define relations between the bivalent variables $x_{id}^k * x_{jd'}^{k'}$ the following two constraints are defined:

$$x_{id}^k + x_{jd'}^{k'} - y_{ijdd'}^{kk'} \leq 1 \qquad \forall i, j = 1, ..., ID; \quad i \neq j; \quad \forall k, k', d, d' = 1, ..., D \tag{9}$$

$$\sum_{kk'dd'} y_{ijdd'}^{kk'} \leq \sum_k x_{id}^k \qquad \forall d = 1, ..., D \tag{10}$$

Each federated data center storage k has a quotas storage space offered to the insourcing/outsourcing intermediate data placement:

$$\sum_{id}^{k \neq d} x_{id}^k \cdot size_i \leq SCF_k \qquad \forall k = 1, ..., D \tag{11}$$

Each data center storage d has a quotas storage space available to local intermediate data placement decision:

$$\sum_{id}^{k = d} x_{id}^k \cdot size_i \leq SCL_d \qquad \forall k = 1, ..., D \tag{12}$$

Table 1. Storage cost federation model: parameters

Parameters	Description
D	Set of federated data centers storage
d, d', k, k'	Are used to designate local data centers storage d, d' and federated data centers storage k, k'
IDL_{id}	A local data center storage d hosting intermediate data i
Dep_{ij}	Intermediate data dependency matrix coefficients as input to the model to designate affinity between set (i and j)
$DCsource_d$	A source data centers matrix which brings local data centers d that generate intermediate data sets
LSC_d	Local storage cost (dollar per GB) of cloud storage provider hosting data center d
OSC_k	Outsourcing storage cost (dollar per GB) of cloud storage provider hosting data center k
$size_i$	Size of intermediate data set i
λ_i^j	Scientific users's tolerance of intermediate data dependencies of i and j which is a binary value: $\lambda_i^j = 0$, indicates scientific users have no tolerance to process i and j independently, and 1 otherwise.
IDi	Set of intermediate data set i
IDN_d	Number of intermediate data set generated and stored temporary in the data center storage d
ITC_k	Insourcing transfer cost proposed by data center storage k
$IOPC_{i,j}$	Cost of I/O-demands (dollar per operation) of intermediate data sets i, j on federated data centers storage
SCF_k	Storage resource quotas offered and shared (GB per month) by each data center k in the storage federation
$SCmax_k$	Maximum storage resource quotas (GB per month) offered and shared by all data center storage federation
SCL_k	Storage resource quotas (GB per month) available by each local data center d
Decision variable	Definition
x_{id}^k	A binary variable, $x_{id}^k = 1$ if intermediate data set i is allocated from data center storage d to outsourced data center storage k, and 0 otherwise.
$y_{ijdd'}^{kk'}$	A binary variable, $y_{ijdd'}^{kk'} = x_{id}^k * x_{jd'}^{k'}$

For all intermediate data set IDi generated from data center source d, the size of IDi placement cannot exceeded the total storage capacity of the federated data center storage:

$$\sum_{id} x_{id}^k \cdot size_i \leq SCmax_k \qquad \forall k = 1, ..., D \qquad (13)$$

Each intermediate data set i are generated from one local data center storage d in a single workflow phase:

$$\sum_d IDL_{id} = 1 \qquad \forall i = 1, ..., ID \tag{14}$$

Intermediate data dependency matrix DEP is symmetric:

$$Dep_{ij} = Dep_{ji} \qquad \forall i, j = 1, ..., ID; \quad \forall Dep_{ij} \in DEP \tag{15}$$

5   Evaluation Results

5.1   Experimental Input Data and Scenario

The assessment concerns the cost optimization of intermediate data dependencies placement. A number of geographical distributed data center storage (in $[2, 10]$) from different cloud provider were drawn randomly to perform the assessment in order to span the optimization space as much as possible. Each outsourcing/insourcing demand is composed of a random number of intermediate data sets dependencies (ranging in $[50 * 50, 1000 * 1000]$) organized in a matrix with $2\,GB$ size per set. The intermediate data sets are affected to their local data center storage in a binary matrix. The insourcing storage monetary is given by Eq. (1), the outsourcing and insourcing transfer cost are fixed (in $[0, 0.001]$), (in $[0.007, 0.02]$) respectively. The affected cost for the end-scientific users monetary and I/O operation on-demands according to the pricing plan such as Amazon S3 [15], Microsoft Azure [16], B2 Cloud Storage Pricing [17] and Google Cloud Storage [18]. The binary value of λ_i^j is set randomly for each new generated intermediate data set.

Since related approaches on intermediate data placement and storage cost optimization can not be directly compared with our exact algorithm as reviewed in Sect. 2, the proposed exact algorithm is validated by comparing with different scenario that not considers federation aspect and capacity-based placement. The data centers storage in non-federation scenario turn in an autonomous way and depend on their own storage space capacity to allocate intermediate data sets.

To elaborate this scenario, a relaxation of the linear program was constructed and consists on eliminating the constraints (5), (7), (8), (11) and (13). The intermediate data placement will not be allocate entirely if the storage space are not available locally, so the rest of unallocated intermediate data will be lost. The outsourcing/insourcing storage cost are obviously not integrated to solve the non-federation algorithm considering just local dependencies. In capacity-based scenario, the data center federation outsource the storage space to allocate the intermediate data sets to the most appropriate federation members only when their own space storage resources are not available. Here, the data center storage selection is done randomly to outsource intermediate data sets without considering the dependencies (constraints (7) and (8)).

We applied the following metrics to analyze the impact of the proposed exact algorithm in the data centers storage federation and comparison scenarios:

1. Total storage cost: This metrics is the objective function computed by Eq. 4.
2. Federation utilization: This metrics shows the intermediate data distribution on a selected data centers storage in the federation. It is defined as the ratio between the amount of storage space used by intermediate data placement (both local and federation members) and the maximum amount of storage space for all intermediate data sets placement.
3. Convergence time of the exact placement algorithm.

5.2   Results and Discussion

In order to evaluate the proposed model and to show the influence of using federated data center storage characteristics which having different storage capacities from cloud provider scenario, we performed a set of experiences with different data sets size. The model evaluation is performed under CPLEX [19] as a MILP optimization solver to solve the objective function.

Figure 3 is the results of minimizing the total storage cost ($) for our model federation and the other two comparison scenarios. The simulations correspond to the different dependency matrix coefficient values which represent workflow phases. A federation of 10 data centers storage cost results aggregation receive outsourcing/insourcing demands of intermediate data sets with different size. It is clear that the most appropriate is the federation results with 27.83% of cost saving regarding to the non-federation scenario and 36.36% compared to the capacity-based scenario while intermediate data size reached 2000 GB (1000 * 2 GB). Indeed, number of dependencies increase with the size of rising intermediate data sets, so the intermediate data dependencies transfer cost is minimized in the federation and this influences the total storage cost of federation results. In addition the I/O operation on-demands cost is minimized when

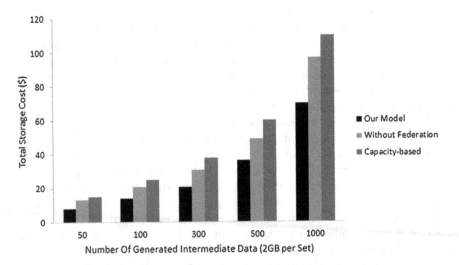

Fig. 3. Minimum total storage cost optimization results

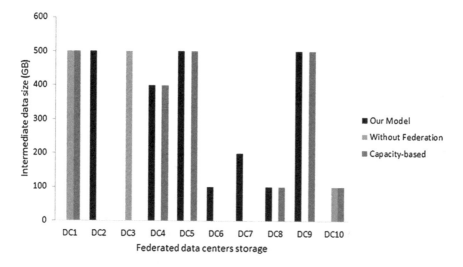

Fig. 4. Intermediate data distribution results in federated data centers storage

there is a dependency tolerance. In the scenario without federation, I/O operation on-demands cost is not minimized, since it considers local dependencies and takes a fixed cost of local data center regardless of the price ranges. The storage cost is high in the capacity-based scenario, so it does not optimize the movement of intermediate data since it allocates them randomly to the different data centers until the capacity is full without taking its dependencies in consideration.

The very important point in the federation is the allocation balancing between the federation members from intermediate data distribution that is illustrated in Fig. 4. We clearly see in the federation scenario that involves 7/10 of data centers storage to allocate intermediate data sets of 2000 GB size compared to 6/10 for the capacity-based scenario and 3/10 only for the scenario without federation regardless of the intermediate data dependencies. As long as the later it participates individually, the allocation and the storage cost will not be optimal through capacity constraint and fixed price. Although the federation is seen not only reduces the storage cost due to its dynamic storage price which is adapted to the intermediate data requirements, but also maintains collaboration among data centers members of the federation for charging and reductions for scientific user community.

The size of intermediate data impact on the federation that is evaluated using 10 data centers storage. The convergence time of the model depends on the size of dependency matrix, however, using a commercial solver like CPLEX, solving time grow to seconds by increasing the size of the matrix. So, the time needed for intermediate data allocation increases to second in Fig. 5 above 360000 intermediate data sets dependencies. The curve is reported for up to 1000000 dependencies for extreme case, this reaches 200 s.

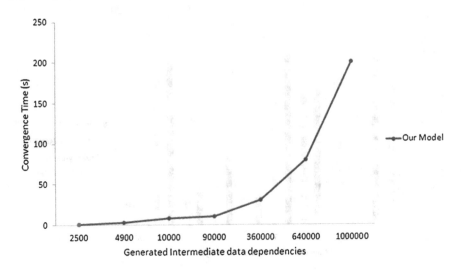

Fig. 5. Convergence time of the exact algorithm

6   Conclusion

The presented work introduced intermediate data storage cost saving solution through federated data centers storage. A MILP model is generated for the allocation problem formulation. The MILP optimization problem takes into account the storage federation characteristics. An exact algorithm is proposed for an optimal allocation taking into account intermediate data dependency, data size and cost saving over a distributed data centers storage. A binary symmetric matrix is defined to represent the dependency for each intermediate data sets and home data center storage hosting matrix are used by our exact algorithm to outsourcing intermediate data storage to the federation. Our proposed model was tested on a set of data sets generated randomly based on values obtained from realistic data. An effective and optimal solution is shown by the proposed exact algorithm and their validation. The algorithm complexity is essentially due to the size of the matrix dependency. So in future work, we plan to improve intermediate data allocation cost by applying data replication and multi data center distance cost. Evenly, to address the convergence time of the exact algorithm, we will proposed a heuristic based intermediate data allocation cost to have a better scalability in a very short duration.

References

1. Dean, J., Ghemawat, S.: MapReduce: simplified data processing on large clusters. Commun. ACM **51**(1), 107–113 (2008)
2. Google Cloud Platform. https://cloud.google.com/bigtable/
3. Workflow Engines. https://wiki.openstack.org/wiki/NovaOrchestration/Workflow Engines

4. Apache Hadoop Core. http://hadoop.apache.org/core
5. Swift-Open Stack. http://docs.openstack.org/developer/swift/
6. Amazon Web Services. https://aws.amazon.com/fr/
7. Kyriazis, D. (ed.): Data Intensive Storage Services for Cloud Environments. IGI Global, Hershey (2013)
8. Yuan, D., Yang, Y., Liu, X., Zhang, G., Chen, J.: A data dependency based strategy for intermediate data storage in scientific cloud workflow systems. Concurrency Comput. Pract. Experience **24**(9), 956–976 (2012)
9. Yuan, D., Yang, Y., Liu, X., Chen, J.: A data placement strategy in scientific cloud workflows. Future Gener. Comput. Syst. **26**(8), 1200–1214 (2010)
10. Zhao, Q., Xiong, C., Zhao, X., Yu, C., Xiao, J.: A data placement strategy for data-intensive scientific workflows in cloud. In: 2015 15th IEEE/ACM International Symposium on Cluster, Cloud and Grid Computing (CCGrid), pp. 928–934. IEEE, May 2015
11. Ruiz-Alvarez, A., Humphrey, M.: A model and decision procedure for data storage in cloud computing. In: 2012 12th IEEE/ACM International Symposium on Cluster, Cloud and Grid Computing (CCGrid), pp. 572–579. IEEE, May 2012
12. Agarwala, S., Jadav, D., Bathen, L.A.: iCostale: adaptive cost optimization for storage clouds. In: IEEE International Conference on Cloud Computing (CLOUD), pp. 436–443. IEEE (2011)
13. Negru, C., Pop, F., Cristea, V.: Cost optimization for data storage in public clouds: a user perspective. In: Proceedings of 13th International Conference on Informatics in Economy (2014)
14. Toosi, A.N., Calheiros, R.N., Thulasiram, R.K., Buyya, R.: Resource provisioning policies to increase IaaS provider's profit in a federated cloud environment. In: 2011 IEEE 13th International Conference on High Performance Computing and Communications (HPCC), pp. 279–287. IEEE, September 2011
15. Tarification Amazon S3. https://aws.amazon.com/fr/s3/pricing/
16. Tarification - Microsoft Azure. https://azure.microsoft.com/fr-fr/pricing/details/storage/
17. B2 Cloud Storage Tarification. https://www.backblaze.com/b2/cloud-storage.html
18. Google Cloud Storage Pricing. https://cloud.google.com/storage/pricing
19. http://ampl.com/products/solvers/solvers-we-sell/cplex/

Tasks Scheduling and Resource Allocation for High Data Management in Scientific Cloud Computing Environment

Esma Insaf Djebbar[(⊠)] and Ghalem Belalem

Department of Computer Science, University of Oran1,
Ahmed Ben Bella, Oran, Algeria
esma.djebbar@gmail.com,
ghalemldz@yahoo.fr

Abstract. Cloud computing refers to the use of computing, platform, software, as a service. It's a form of utility computing where the customer need not own the necessary infrastructure and pay for only what they use. Computing resources are delivered as virtual machines. In such a scenario, data management in virtual machines in Cloud Computing is a new challenge and task scheduling algorithms play an important role where the aim is to schedule the tasks effectively so as to reduce the turnaround time and improve resource utilization and Data Management.

In this work, we propose two strategies for task scheduling and resource allocation for high data in Cloud computing. The main objective is to improve data management in virtual machine in Cloud computing and optimize the total execution time of all tasks.

Keywords: Task scheduling · Resource allocation · High data management · Scientific cloud computing

1 Introduction

Cloud computing can be best described as a highly automated, readily scalable, on-demand computing platform of virtually unlimited processing, storage and ubiquitous connectivity, always available to carry out a task of any size and charged based on usage. While early in its evolution, Cloud computing is fast becoming as pervasive a platform as the internet [11].

Cloud computing is the large-scale Data center resources which are more concentrated [1]. In addition, virtualization technology hides the heterogeneity of the resources in Cloud computing [2], Cloud computing is user-oriented design which provides varied services to meet the needs of different users. It is more commercialized, and the resources in Cloud computing are packed into virtual resources by using virtualization technology [3]. This causes its resource allocation process, the interaction with user tasks and so on are different with grid computation.

Cloud computing adoption continues to gain momentum across a broad range of industries including financial services. Once organizations manage to filter through the

© Springer International Publishing AG 2016
S. Boumerdassi et al. (Eds.): MSPN 2016, LNCS 10026, pp. 16–27, 2016.
DOI: 10.1007/978-3-319-50463-6_2

noise surrounding the cloud, there are actually some very pragmatic ways in which the IT organizations of banks, insurers and similar institutions can leverage Cloud computing to directly benefit their daily operations and most significantly, have an impact on the business bottom line. These gains can be achieved without incurring large capital expenditure or exposing sensitive business data.

Cloud task scheduling and virtual machine (VM) resource allocation optimization is an important, challenging and core component in the cloud application services and Cloud computing systems. Scheduling refers to the appropriate assignment of tasks to the resources available like CPU, memory and storage, such that there is a maximum utilization of resources. Efficient scheduling is a necessary for both cloud service requesters as well as providers.

In this work, we propose two strategies for tasks scheduling and resource allocation. The first strategyy for tasks scheduling and resources allocation is based on a deadline, length of cloudlets and the speed of execution of the virtual machine. The second strategy of task scheduling and resource allocation uses tree based data structure called Virtual Machine Tree (VMT) for efficient execution of tasks.

The remaining parts of this paper are organized as follows. The next section briefly describes the generality of tasks scheduling in a Cloud computing environment. In Sect. 3, the related works are presented for a task scheduling and resource allocation in a Cloud computing environment. A proposed model is illustrated in Sect. 4 and experimentation and results is given in Sect. 5. Section 6 concludes the paper and discusses future research directions.

2 Problem of Task Scheduling

Task scheduling algorithm is a method by which tasks are matched, or allocated to Data center resources. Due to conflicting scheduling objectives generally no absolutely perfect scheduling algorithm exists. A good scheduler implements a suitable compromise, or applies combination of scheduling algorithms according to different applications [12]. A problem can be solved in seconds, hours or even years depending on the algorithm applied. The efficiency of an algorithm is evaluated by the amount of time necessary to execute it. The execution time of an algorithm is stated as a time complexity function relating the input. There are several kinds of time complexity algorithms that appear in the literature. If a problem has a polynomial time algorithm, the problem is tractable, feasible, efficient or fast enough to be executed on a computational machine. In computational complexity theory, set of problems can be treated as complexity class based on a certain resource.

Class NP is the set of decision problems that are solvable on a nondeterministic Turing machine in polynomial time, but a candidate solution of the problem of Class NP can be confirmed by a polynomial time algorithm, which means that the problem can be verified quickly.

Class NP-complete is the set of decision problems, to which all other NP problems can be polynomial transformable, and a NP-complete problem must be in class NP. Generally speaking, NP-complete problems are more difficult than NP problems.

Class NP-hard is the set of optimization problems, to which all NP problems can be polynomial transformable, but a NP-hard problem is not necessarily in class NP.

Task scheduling problem is the problem of matching tasks to different sets of resources which is formally expressed as a triple (T, S, O) where 'T' is the set of tasks, each of which is an instance of problem, the set of feasible solutions is 'S' and the objective of the problem is 'O'.

Scheduling problem can be further classified into two types as optimization problem and decision problem based on objective O. An optimization problem requires finding the best solution among all the feasible solutions in set S. Different from optimization; the aim of decision problem is relatively easy. For a specified feasible solution s ∈ S, problem needs a positive or negative answer to whether the objective is achieved. Clearly, optimization problem is harder than decision problem.

Scheduling theory for Cloud computing is receiving growing attention with increase in cloud popularity. In general, scheduling is the process of mapping tasks to available resources on the basis of tasks' characteristics and requirements. It is an important aspect in efficient working of cloud as various task parameters need to be taken into account for appropriate scheduling. The available resources should be utilized efficiently without affecting the service parameters of cloud. Scheduling process in cloud can be generalized into three stages namely:

a. **Resource discovering and filtering:** Datacenter Broker discovers the resources present in the network system and collects status information related to them.
b. **Resource selection:** Target resource is selected based on certain parameters of task and resource.
c. **Task submission:** Task is submitted to resource selected. This is deciding stage.

3 Related Works

The difficulty of task scheduling in distributed computing system is to handle dependent or independent tasks is a well studied area. In this section, we explained different existing task scheduling methods in a heterogeneous computing environment. By using dynamic allocation methods applied on large sets of real-world applications that are able to be formulated in a way which allows for deterministic execution. But the dynamic technique doesn't have any prior knowledge about tasks to be executing compare to static techniques.

The task scheduling algorithms currently prevalent in clouds are summarized in Table 1. All the algorithms are implemented in cloud computing environment.

Target resources in a cloud environment can be selected in various ways. The selection of resources can be either random, round robin, greedy (resource processing power and waiting time based) or by any other means. The selection of jobs to be scheduled can be based on FCFS, SJF, priority based, coarse grained task grouping etc. Scheduling algorithm selects job to be executed and the corresponding resource where

Table 1. The related works in the literature

N	Scheduling algorithms	Scheduling parameters	Skeleton outline	Tools
1	Cloud task and virtual machine allocation strategy [4]	The total execution time of all tasks	Use Speed of VM, cloudlets length and load balancing is taken in to account	Cloudsim
2	Optimal scheduling of computational task [5]	The execution time	Use Tree based data structure called Virtual Machine Tree (VMT) for efficient execution of tasks	Cloudsim
3	A deadline scheduler for jobs [6]	The missed deadlines, execution time and cost	Cloud Least Laxity First (CLLF), minimizes the extra-cost implied from tasks that are executed over a cloud setting by ordering each of which using its laxity and locality	Cloudsim
4	Job scheduling algorithm based on Berger [7]	The execution time, user satisfaction and CPU number	The algorithm establishes dual fairness constraint	Cloudsim
5	Online optimization for scheduling preemptable tasks on IaaS cloud systems [8]	The computational power, execution time, bandwidth	Algorithms adjust the resource allocation dynamically based on the updated information of the actual task executions	Simulation environment
6	A priority based job scheduling algorithm [9]	The complexity, consistency, makespan	The proposed algorithm is based on multiple criteria decision making model	Cloud environment
7	Performance and cost evaluation of Gang Scheduling in a Cloud Computing system with job migrations and starvation handling	The response time and Bounded Slowdown	The study takes into consideration both performance and cost while integrating mechanisms for job migration and handling of job starvation	Cloud environment

the job will be executed. As each selection strategy is having certain flaws work could be done in this direction to extract the advantageous points of these algorithms and come up with a better solution that tries to minimize the drawbacks of resultant algorithm.

New scheduling strategy need to be proposed to overcome the problem posed by network properties and user requirements. The new strategies may use some of the conventional scheduling concepts to merge them with some network and requirement aware strategies to provide solution for better and more efficient task scheduling.

4 Proposed Strategy of Task Scheduling

In this section, two strategies of task scheduling and resources allocation are presented:

4.1 The First Strategy

We propose a strategy for tasks scheduling and resources allocation based on a deadline, length of cloudlets and the speed of execution of the virtual machine. Our proposition are different from [9] because we add in the algorithm, in the second step,

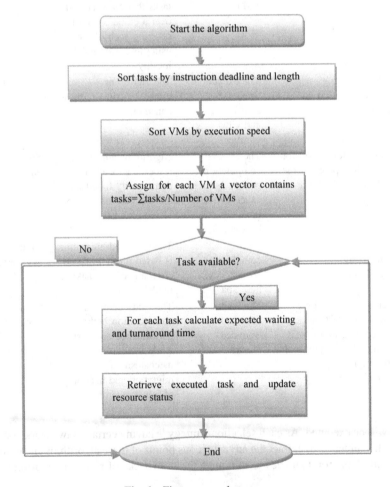

Fig. 1. First proposed strategy

a division of number of cloudlets by the number of virtual machines to minimize an average time execution of all tasks. The major process lines of strategy are as follows, and the flow chart is shown in Fig. 1.

The different steps of the algorithm are as follows:

Step 1: sort the cloud task by instruction deadline and length in ascending order;
Step 2: sort the virtual machine by execution speed in ascending order;
Step 3: Assign for each VM a vector, a number of cases equals to M (number of cloudlets) devise by N (number of VM); so that the first group of the first tasks are executed by the first VM, the second one are executed by the second VM,...

4.2   The Second Strategy

The second strategy of task scheduling and resource allocation uses tree based data structure called Virtual Machine Tree (VMT) for efficient execution of tasks. Our algorithm is an amelioration of [10] but it provides a better load balancing.

A Virtual Machine Tree (VMT) is a binary tree with N nodes. Each node represents a Virtual Machine containing Virtual Machine Id and MIPS. N represents total number of computational specific Virtual Machines in a cloud. The special property of VMT is that node value (MIPS) at level L is greater than or equal to node value at level L + 1 where L $\geq$ 0. Each node contains zero, one or two child nodes. A node with no child node is called as a leaf node and the node with child nodes is referred as internal nodes. Consider a 5 computational specific Virtual Machines represented by their Id and MIPS as V = {{0, 250}, {1, 1000}, {2, 250}, {3, 500}, {4, 250}}. Figure below shows the VMT. The VMT is constructed based on the prioritized order of Virtual Machines from left to right, such that Virtual Machine with highest MIPS becomes the root (Fig. 2).

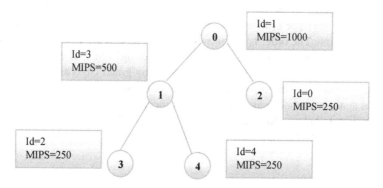

Fig. 2. An example of a Virtual Machine Tree (VMT)

Here VMT with the root node representing the Virtual Machine with Id 1 and MIPS 1000. The root node has two children. The left child node represents the Virtual Machine with Id 3 and MIPS 500. The right child node represents the Virtual Machine with Id 0 and MIPS 250. Similarly node which represents the Virtual Machine with Id 3 and MIPS 500 has 2 children. The left child of this node represents the Virtual Machines with Id 2 and MIPS 250, right child represents the Virtual Machine with Id 4 and MIPS 250 respectively.

Here we present a grouping mechanism for the set of task submitted to the cloud. Let T COUNT be the total number of tasks submitted and L COUNT be the total number of leaf nodes in VMT. The total number of groups G COUNT for the submitted tasks are computed as follows: G COUNT = L COUNT.

If VMT constructed with 5 Virtual Machines, then total number of group is the number of level and it's equal to 3 in our example. The number of tasks in each group G is computed as follows. G = Number of level.

Each group contains the maximum number of tasks that their sum does not exceed a value which is calculated by the following formula, and each group is assignaded in this level, the first level in the top level (root), the second group in the second level, and the last one in the third level.

$$\sum length\ of\ tasks * VM\ Mips\ in\ level \Big/ \sum Mips\ of\ VMs$$

Consider 12 tasks represented by their Id and size as G = {{0, 20000}, {1, 20000}, {2, 20000}, {3, 10000}, {4, 10000}, {5, 20000}, {6, 10000}, {7, 20000}, {8, 10000}, {9, 10000}, {10, 20000}, {11, 10000}}.

After prioritizing and grouping, each group contains following tasks.

G1 = {{0, 20000}, {1, 20000}, {2, 20000}, {5, 20000}}
G2 = {{7, 20000}, {10, 20000}, {3, 10000}, {4, 10000}}
G3 = {{6, 10000}, {8, 10000}, {9, 10000}, {11, 10000}}

Once the grouping of the tasks are done, suitable Virtual Machines are selected for execution from VMT. The tasks in each group is selected sequentially and submitted to the Virtual Machine. The order is as follows. The first task in the group G1 is executed by the Virtual Machine represented by the root node of the VMT. The second task will be executed by its child, third task will be executed by grand child and so on. Once it reaches the Virtual Machine represented by the leaf node, the next task will be submitted once again to root node and so on. Same procedure is repeated for all the tasks in each group. Figure below shows the VMT for 5 Virtual Machines and total number of groups formed for the 12 submitted tasks (Fig. 3).

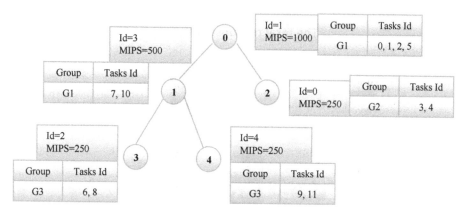

Fig. 3. The result of Tasks execution in the VMT tree

Here total number of tasks submitted will be in 3 groups namely G1, G2 and G3 respectively. Tasks with Id 0, 2, 5, 7 will be in group G1, tasks with Id 10, 1, 3, 4 will be in group G2 and tasks with Id 6, 8, 9, 11 will be in group G3 respectively.

5   Experimentation Results

5.1   The Response Time for the First Strategy

The experiments are conducted on a simulated Cloud environment provided by CloudSim. The speed of each processing element is expressed in MIPS (Million Instructions Per Second) and the length of each cloudlet is expressed as the number of instructions to be executed. The simulation environment consists of two Data Center with two hosts having two Processing Elements respectively. Each Processing Element is assigned varying computing power (varying MIPS). The algorithms are tested by varying the number of cloudlets from 10 to 50 and also varying the length of cloudlets. Also, the number of VMs used to execute the cloudlets, are varied accordingly. The overall response time to execute the cloudlets is used as the metric to evaluate the performance of the first strategy. The results are shown in Table 2 and Fig. 4.

Table 2. The total response time results of execution tasks

	Time shared	Space shared	Proposed first strategy
Response time 10 Cloudlets/s	734,92	840,75	646,63
Response time 30 Cloudlets/s	3185,15	2204,34	2032,39
Response time 50 Cloudlets/s	8959	3776,8	3548,35

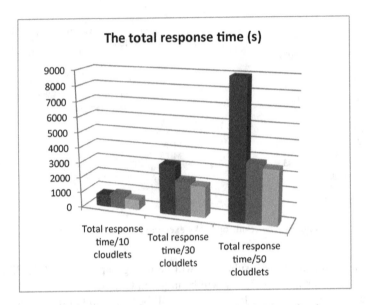

Fig. 4. The total response time results of execution of tasks.

The following Fig. 5 show the gain obtained:

It has been observed that, for smaller number of tasks, all the three algorithms exhibit more or less similar performance since the length of the queued cloudlets is less. But as shown in Table 2 and Fig. 4, as the number of tasks increase, the first strategy exhibits better performance when compared to Space shared and Time shared since longer tasks complete faster thereby reducing the response time.

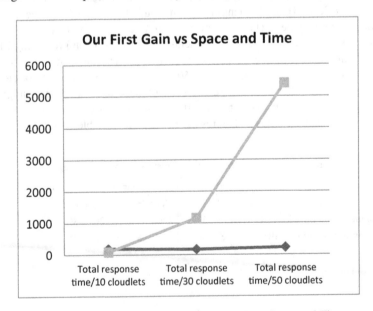

Fig. 5. The gain of the first strategy obtained vs. Space and Time.

5.2   The Response Time for the Second Strategy

For the second strategy, the experiments are conducted on a simulated Cloud environment provided by CloudSim. The speed of each processing element is expressed in MIPS (Million Instructions Per Second) and the length of each cloudlet is expressed as the number of instructions to be executed. The simulation environment consists of two Data Center with two hosts having two Processing Elements respectively. Each Processing Element is assigned varying computing power (varying MIPS). The algorithms are tested by varying the number of cloudlets from 10 to 50 and also varying the length of cloudlets. Also, the number of VMs used to execute the cloudlets, are varied accordingly. The overall response time to execute the cloudlets is used as the metric to evaluate the performance of the first strategy. The results are shown in Table 3 and Fig. 6.

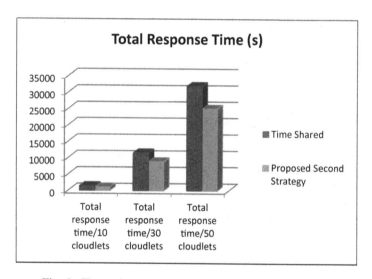

Fig. 6. The total response time results of execution of tasks.

The following Fig. 7 show the gain obtained:

It has been observed that, for smaller number of tasks, all the three algorithms exhibit more or less similar performance since the length of the queued cloudlets is less. But as shown in Table 3 and Fig. 6, as the number of tasks increase, the second strategy exhibits better performance when compared to Time shared since longer tasks complete faster thereby reducing the response time.

Table 3. The total response time results of execution tasks.

	Time shared	Proposed second strategy
Response time 10 Cloudlets/s	1334,94	980
Response time 30 Cloudlets/s	11475,05	8819,98
Response time 50 Cloudlets/s	31875,7	24959,88

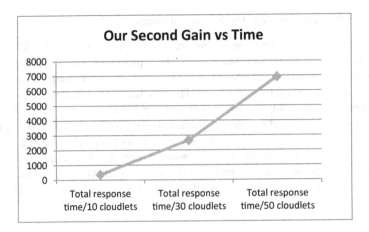

Fig. 7. The gain of the second strategy obtained vs. Time.

The two strategies can provide a better response time, a minimizing waiting time of tasks and better load balancing.

6 Conclusion and Perspectives

In this paper, we proposed two effective scheduling strategies and allocation resources in the environment of computing clouds. The first strategy is based on the length of tasks and the speed of execution of Virtual Machines, the second is based on VMT (Virtual Machine Tree).

The two strategies are tested in the CloudSim simulator and compared with Space Shared and Time Shared politics. The strategies can provide a better response time, a minimizing waiting time of tasks and better load balancing.

For a continuation of our work, we think to propose a third strategy of task scheduling and resource allocation takes into account the pre-emption of tasks to ensure better energy consumption and in the second way, The future work may group the cost based tasks before resource allocation according to resource capacity to reduce the communication overhead.

References

1. Peng, L.: The definition of cloud computing and characteristics. http://www.chinacloud.cn/. 2009-2-25
2. Sutherland, I.E.: A future market in computer time. Commun. ACM **11**(6), 449–451 (1968)
3. Ferguson, D., Yemini, Y., Nikolaou, C.: Microeconomic for load balancing in distributed computer Systems. In: Proceeding of the Eighth International Conference on Distributed Systems. San Jose, pp. 491–499. IEEE Press (1988)

4. Xu, X., Hu, H., Hu, N., Ying, W.: Cloud task and virtual machine allocation strategy in cloud computing environment. In: Lei, J., Wang, F.L., Li, M., Luo, Y. (eds.) NCIS 2012. CCIS, vol. 345, pp. 113–120. Springer, Heidelberg (2012)
5. Achar, R., Thilagam, P.S., Shwetha, D., et al.: Optimal scheduling of computational task in cloud using virtual machine tree. In: 2012 Third International Conference on Emerging Applications of Information Technology (EAIT), pp. 143–146 (2012)
6. Perret, Q., Charlemagne, G., Sotiriadis, S., Bessis, N.: A deadline scheduler for jobs in distributed systems. In: 2013 27th International Conference on Advanced Information Networking and Applications Workshops (WAINA), pp. 757–764 (2013)
7. Baomin, X., Zhao, C., Enzhao, H., Bin, H.: Job scheduling algorithm based on Berger model in cloud environment. Adv. Eng. Softw. **42**, 419–425 (2011)
8. Li, J., Qiu, M., Ming, Z., Quan, G., Qin, X., Gu, Z.: Online optimization for scheduling preemptable tasks on IaaS cloud systems. J. Parallel Distrib. Comput. **72**(5), 666–677 (2012)
9. Ghanbaria, S., Othmana, M.: A priority based job scheduling algorithm in cloud computing. In: ICASCE 2012, pp. 778–785 (2012)
10. Moschakis, I.A., Karatza, H.D.: Performance and cost evaluation of Gang Scheduling in a Cloud Computing system with job migrations and starvation handling. In: IEEE Symposium on Computers and Communications (ISCC) (2012) and 2011 IEEE Symposium on Computers and Communications, pp. 418–423 (2011)
11. Sharma, A.: Data management and deployment of cloud applications in financial institutions and its adoption challenges. Int. J. Sci. Technology Res. **1**(1), 1–7 (2012)
12. Djebbar, E.I., Belalem, G.: Optimization of tasks scheduling by an efficacy data placement and replication in cloud computing. In: Aversa, R., Kołodziej, J., Zhang, J., Amato, F., Fortino, G. (eds.) ICA3PP 2013, Part II. LNCS, vol. 8286, pp. 22–29. Springer, Heidelberg (2013). doi:10.1007/978-3-319-03889-6_3

Autonomous Intercloud Reference Architecture Driven by Interoperability

Yosra Abassi[✉], Cherif Ghazel, and Leila Saidane

National School of Computer Sciences, University of Manouba,
Manouba, Tunisia
yosra.abassi@gmail.com, cherif.ghazel@email.ati.tn,
leila.saidane@ensi.rnu.tn

Abstract. Cloud computing is on the hype nowadays, new offering appears frequently and the demand on cloud services is growing in an exponential rate. In such dynamic and competitive market, working in silos is not a successful strategy. Therefore, single providers should integrate their resources and move up to a new level of cooperation and interconnection. This federation would bring in added value in terms of QoS, security, economics, etc. but also more complexity regarding federation management and how interoperability between disparate providers would be established. In this paper, we will present our federation concept through a layered autonomous intercloud reference architecture driven by interoperability.

Keywords: Intercloud · Cloud federation · Interoperability · Autonomy · Reference architecture

1 Introduction

Cloud computing is a set of elastic virtualized resources that can be accessible over a network from ubiquitous devices using a thin client interface such as a web browser in a pay-as-you-go fashion. In this computing model, customers do not have to own or manage a hardware infrastructure, instead they send requests to one or more cloud providers that perform the work on their behalf and return the result following a client/server model.

The concept underlying cloud computing is not new; it dates back to 1961 when the approach of "utility computing" has been introduced for the first time. The idea consisted of a future platform allowing computing resources to be sold as the fifth public utility. Through the 1960's and 70's there has been a popularity of "time-sharing" computing model that suffered latter from lack of powerful software, hardware and networking technologies. In the 1990's we have seen the birth of "grid computing" which is a distributed architecture of large numbers of connected computers with a purpose of solving complex problems. Through the early new millennium, the proliferation of technological advances combined with the change in mindset has enabled a wide spread of Internet-based services called "cloud computing". The latter is not a revolution but rather the outcome of technological evolutions over many years.

© Springer International Publishing AG 2016
S. Boumerdassi et al. (Eds.): MSPN 2016, LNCS 10026, pp. 28–37, 2016.
DOI: 10.1007/978-3-319-50463-6_3

Fig. 1. Intercloud overview

It improves old concepts inherited from previous generations and addresses new challenges that was not supported before.

However, a single cloud provider is unable to meet the needs of customers dispersed all over the world especially in case of a sudden workload spike. Further, relying only on a single cloud provider could bring the danger of service unavailability and vendor lock-in. To deal with these limitations, cloud providers should integrate their datacenters and interoperate with each other in a transparent federated structure as it was the case with Internet and telecom systems. Thus, an intercloud infrastructure would be the next step after the actual fragmented cloud market. The federation process would be complicated by a number of factors such as security, QoS, semantics, management, networking, open standards, etc. as illustrated in Fig. 1. Therefore, an intercloud reference architecture has to be properly designed with these requirements in mind so that both providers and customers could gain the most out of it.

The next section will introduce some definitions required to understand the concept of intercloud and Sect. 3 will explain why we need this federated system. In Sect. 4, we will enumerate some previous researchers on the same subject. Our intercloud reference architecture would be presented in Sect. 5. In the last section, we will close with some conclusions and future steps.

2 Terminolgy

Since its advent there was not a unanimous consent on a uniform definition of cloud computing. The situation is even worse when talking about intercloud. Moreover, there is a shortage of literature dealing with this subject due to its novelty however, an interesting discussion could be found in [1]. Intercloud is a broad term and it encompasses two main categories: multi-cloud (cf. Fig. 2a) also known as sky

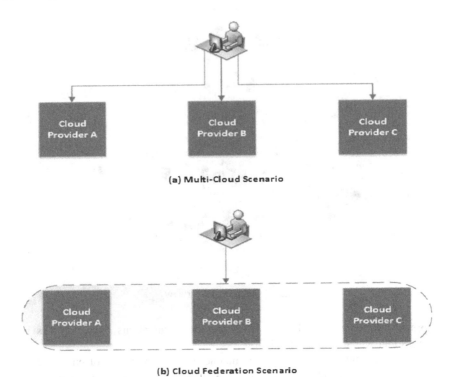

(a) Multi-Cloud Scenario

(b) Cloud Federation Scenario

Fig. 2. Intercloud scenarios

computing and cloud federation. In multi-cloud environment providers are unaware yet unwilling to integrate their systems. Hence, the biggest part of the job would be done by customers using brokering services and libraries in order to enable the aggregation of these independent clouds and to facilitate the migration from one provider to another. Multi-cloud is therefore considered more as a cloud portability mechanism than a real intercloud architecture [2]. On the other hand a cloud federation (cf. Fig. 2b) is the result of a volunteer cooperation of multiple clouds to transparently provide end users with services of better quality. Therefore, members of the federation should play the interoperability card to mask the difference between them and to act as a single system. The burden of federation management would move from the client side to the cloud side.

In a complex distributed system like intercloud, interoperability is a quality of prime importance. Basically, interoperability means the ability of two or more systems (public cloud, private cloud, on-premises) to communicate and seamlessly exchange data through understood interfaces in order to provide customers with coherent and more efficient services that fulfill the expected SLA. Broadly speaking, portability is closely tied to interoperability. It designates the capability of data and applications to effortlessly migrate from one environment to another despite the differences between them. In the same context we can state the definition of cloud integration which refers to the combination of multiple heterogeneous systems into one, eventually more valuable, unit.

Cross-cloud API is a cornerstone of interoperability in intercloud architecture. It offers a common and standardized way to communicate with various proprietary APIs used by different providers [3]. Furthermore, in [4] the writer advocates that APIs should go beyond their traditional functionalities of unified access to CPU, storage and appliance resources to support the orchestration of monitoring, data replication and load balancing. More sophisticated strategies can be adopted for more efficient inter-operability process such as an abstraction layer [5] or a brokering solution [6]. Regardless of the used technique, the key engine of interoperability is semantics that enforce the comprehension and meaningful exchange of information among different components of federation. It can be achieved either by referring to a common semantic model or by using translation and mapping procedures.

3 Motivation

Depending on a single cloud provider will sooner or later raise concerns related to vendor lock-in and services availability. Nowadays many organizations are reluctant to move their critical data and applications to a specific cloud provider due to its underlying proprietary technologies. Moreover it can be confusing to select the most appropriate candidate that satisfy the organization's need in a dynamic always changing market. Another often observed problem is the service unavailability caused by either a technical outage or natural disaster which prevent users from accessing distant computing resources. Furthermore, the fact that one cloud provider can't install datacenters all over the world restricts the locations where sensitive data of users can be stored and that may lead to legal and regulatory issues. This is where the importance of cloud federation comes into play by offering more choice for end users to take the advantage of the best-of-breed solution.

On the other side, cloud providers are always looking to maximize their profit and to optimize the utilization of their resources. Unfortunately that is not a trivial task due to the unforeseen workload. In practice, a cloud provider alternates between two sit-uations: a lean period when there is an underutilization of resources thus resources in excess could be rented to minimize the expenses and a peak period when resources are insufficient to fulfill the users' requests hence the provider should migrate a part of the workload to another provider otherwise the promised SLA would be violated. SLA could be also affected by geographical dispersion of customers which affect the latency and as a consequence the quality of delivered services. In order to overcome these shortcomings cloud suppliers should find the right formula of cooperation within a single federated structure to offer more value-added services.

4 Related Works

In this section, we briefly review some of the main previous research projects focusing on the provision of cloud services in intercloud systems.

The NIST Cloud Reference Architecture [7] identifies five major participants namely the Cloud Consumer, Cloud Provider, Cloud Broker, Cloud Auditor and Cloud

Carrier. The Cloud Consumer can directly request services from Cloud Provider or via the Cloud Broker that help to deal with the complexity of cloud service offerings and may generate value-added services. The Cloud Auditor conducts the independent performance and security monitoring of cloud services and the Cloud Carrier is responsible for the communication and data exchange functionalities. Furthermore, the proposed architecture describes the architectural component for business support and cloud services orchestration, management and security.

Intercloud Architecture for Interoperability and Integration [8] is built upon existing standards particularly the NIST Cloud Computing Reference Architecture. It extends their efforts by defining four complementary components: the Multilayer cloud Service Model (CSM) for vertical cloud services interaction, integration and compatibility, the InterCloud Control and Management Plane (ICCMP) for Intercloud infrastructure and application control and management, the InterCloud Federation Framework (ICFF) for independent clouds and related infrastructure component federation and the InterCloud Operation Framework (ICOF) for multi-provider infrastructure operation.

Cloud4SOA Reference Architecture [9] combines three fundamental computing paradigms namely cloud computing, Service Oriented Architecture (SOA) and light-weight semantics in a layered architecture that targets PaaS users. Other than the SOA and semantic-layers, it presents a governance layer for business centric focus. More-over, it incorporates a front-end layer to support the access of clients to available functionalities through widgetized services and persistency layer that consists of a repository to store semantic and syntactic data. Finally, a harmonized API is used to unify the access to heterogeneous cloud providers.

The Cloud Computing Open Architecture (CCOA) [10] used an SOA approach combined with virtualization technology to form an integrated cloud ecosystem. The definition of relationship between cloud vendors, cloud partners and cloud clients revolves around seven principles namely an integrated ecosystem management, cloud infrastructure virtualization, common reusable services through service-orientation, extensible provisioning and subscription, configurable cloud offerings, unified information and exchange framework and more importantly cloud quality and governance.

The cloud computing architecture based on SOA discussed in [11] presents a unified approach for seamless migration between different cloud providers. The resource layer is covered by the package layer to mask the heterogeneity of various cloud offering. The link layer is based on ESB technology to offer routing, management, monitoring, security, etc. functionalities. The layer just above is the service process layer that is responsible for service publishing, ranking, and dynamic SLA negotiation. The architecture presents also a display layer to enable a common service access.

The aforementioned references architectures presented the core elements to build an intercloud framework. However, there are many open issues regarding supervision, knowledge and autonomy. To counter these problems we advocate our architecture specifically designed with such requirements in mind. Special attention would be devoted to interoperability, supervision and SLA aspects.

5   InterCloud Reference Architecture

An intercloud reference architecture is done by putting together federation components to have a high-level vision of the system.

At the top, the frontend interface acts as a bridge between the customers and the providers by affording them with a simple user-friendly dashboard to facilitate and centralize the access to different cloud offerings. At the bottom, we find the cloud providers that will cooperate to enable the federation. Each participant should contribute by some resources and should present a minimal service level. In between lies most of the core of our intercloud architecture organized in eight horizontal and vertical layers as shown in Fig. 3.

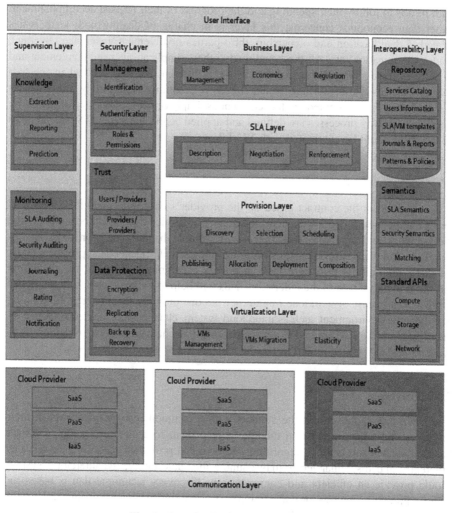

Fig. 3.   Intercloud reference architecture

Each layer represents one of the main functions of the federation and is composed of a number of modules needed to perform these roles.

5.1 Business

The federation acts as marketplace for providers and customers to securely sell and buy computing services. Each provider has his own way to price services for end users and to charge other providers for his resources rent. Moreover, prices are dynamic and could change periodically without a previous notification in most of time. Therefore, the federation should present an efficient accounting and billing mechanism to deal with these issues. This approach should also take into consideration the fact that the number of providers contributing to fulfill the users' requests and their performance vary over the time. On the other side, the intercloud system should grant customers with means to follow their spending and the progress of their request accomplishment. Other than economics concerns, this layer is in charge of dealing with legal issues related to data location and ownership, third-parties involvement, insurance coverage, contract termination, etc. All these and many other complications need convenient and clear rules to regulate the relationship between different parties involved in the federation. Business Process Management (BPM) is becoming increasingly important, as the federation is committed to the continuous improvement of its performance by managing the way processes are defined and applied across the system. Some subtasks of the BPM could be automated with Workflows.

5.2 SLA

It is negotiated in the contract based on the provider's description and the customers' needs and budget. It defines the expected quality of service that would be supplied and the penalties and compensations in case of contract breach. Each provider has its definition and metrics for SLA thus bringing all these heterogeneous approaches under the same umbrella is a thought task and needs a common agreement on a minimal threshold to participate in the federation. This will help to keep the promised quality of service even in case of workload migration from one provider to another. Additionally, in a dynamic environment such as intercloud system, changes and accidents happen all the time, therefore it's required to set a supervision mechanism to detect or better yet to prevent the SLA violation and to take the adequate corrective measures.

5.3 Provision

The federation should afford the vendors with a common repository to publish their resources and with a prediction module to help them forecast the coming workload based on statics and learning algorithms. Known that providers describe their services differently and dissimilar terms may connote the same entity, the intercloud system should provide an efficient and comprehensive discovery technique that is based on many parameters such as provider's QoS, its state and its votes (clients rating and platform evaluation). Then, from the discovered services, the federation should optimally and as automatically as possible select where to deploy the users' data and applications. Further, this step will deliver a list of recommended similar services to choose from in case of workload migration or a provider outage. Next, the order of task execution will be decided based on application and data dependencies and using different algorithms and heuristics. Afterwards, the selected providers will allocate their resources to users' requests in a way that maximize the exploitation of their resources and so their profit without violating the SLA. The federation may also bring together single services from different providers into one composite unit to respond to some specific request or to offer more valuable and efficient services.

5.4   Virtualization

Cloud providers are delivering their services in a shared virtualized trend. The fact that they are using specific configurations and different image formats make it hard for them to cooperate and aggregate their resources to empower the workload traffic between them. To solve this problem the federation will either enforce a unified platform-independent virtual machine format with standardized configuration or more concretely establish a procedure for packaging and live migration of virtual machines from one environment to another. Besides the federation needs a module to manage the lifecycle of virtual machines from creation to destruction and another one to manage the resources elasticity. The latter will adjust the allocated resources to the variation in demand by adding or minimizing the number of used VMs.

5.5   Security

It is one of the most pressing issues in respect with cloud computing let alone with an intercloud system. The main identified concerns are identity management, trust and data protection. The identity management aims to control the access to the intercloud through efficient authentication and authorization procedures that balance the trade-off between the simplicity through a single account for all the federation and the complexity to guarantee an efficient protection from malicious acts. By a user we mean an individual customer, an organization that comprises many users or a cloud computing. Trust is two-fold: a trust between the users and the providers and a trust between the providers themselves based on the performance evaluation delivered by the monitoring layer. Both the identity management and the trust could be established in many levels and managed by a certified third party. The security layer encompasses also a data protection module to prevent users' data and applications from being attacked or lost. To do so the federation should adopt a strong and efficient encryption algorithms and data replication and recovery mechanisms.

5.6   Supervision

Establishing a global supervision process is a paramount for any federated system and intercloud is no exception. The main functionality of this layer is gathering information from all the federation principally the outcomes of SLA and security audit that will be than stored in state warehouse. The monitoring platform should be designed to work in an autonomic fashion and to take into account the distribution that characterizes the intercloud environment. Additionally, it should be equipped with a log process and an appropriate notification and messaging tool to alert the users as well as the providers about the changes and events that occur in the system. Another module that can be offered by this layer is the rating functionality that allow users to give their feedback about the quality of proposed services and the federation to evaluate the performance of participants based on how much SLA is kept. Based on the output of the monitoring activity, a knowledge engine will apply specific machine learning algorithms to categorize and synthetize the raw data and extract useful information and patterns in order to prepare statics and reports about the current and future state of the federation based on previous experiences. These reports are used by most of the other modules such as SLA enforcement, service discovery and selection, trust, etc. That will help to timely detect and react to changes and troubles as well as to predict and plan for upcoming steps.

5.7 Interoperability

This layer plays a catalyst role in the entire system and it operate along three main axes: shared repository, semantic models and open APIs. An intercloud repository is a common place to store the service catalog, the users' information, the monitoring output, the ready to use SLA templates and virtual machine images. It should be designed in an integrated but not centralized approach due to scalability and reliability issues. Besides, the federation needs semantics to overpass the heterogeneity and eventual conflicts in SLA syntax and security policies and to facilitate the service discovery and data portability using formal language such as ontologies. Conjointly the intercloud system requires open and vendor-independent APIs to cover the proprietary providers' native APIs and hence to enable a unified access to available computing, storage and networking resources.

5.8 Communication

In order to facilitate the communication between the different parts of the federation, a scalable non-central component for messages distribution should be designed. This entity would be aimed to support the exchange of increasing flow of data across the whole system in minimal delay.

6 Conclusion

Comparing to traditional on-premises systems cloud computing helps to maximize profit and optimize the resources exploitation. What is more our ability to make things much better by combining the effort of multiple cloud providers to increase users' choice and allow more players and thereby enhance the competition and the quality of offered services. In this regard, we present an autonomous intercloud reference architecture based on interoperability and auto-supervision mechanisms. Our contribution consists of proposing various modules to enable an efficient and comprehensive communication between different participants. By the advocated model, we paint our vision to the future of cloud computing as a smart integrated system. Our upcoming work regarding this reference architecture would be the actual definition and description of components that would constitute the cloud federation platform.

References

1. Grozev, N., Buyya, R.: Inter-Cloud Architectures and Application Brokering: Taxonomy and Survey, pp. 369–390. Wiley, Hoboken (2012)
2. Bernstein, D., Demchenko, Y.: The IEEE intercloud testbed – creating the global cloud of clouds. In: IEEE 5th International Conference on Cloud Computing Technology and Science (CloudCom), Bristol, vol. 2, pp. 45–50, December 2013
3. Petcu, D., Craciun, C., Rak, M.: Towards a cross platform cloud API - components for cloud federation. In: 1st International Conference on Cloud Computing and Services Science, Netherlands, May 2011
4. Ranjan, R.: The cloud interoperability challenge. IEEE Cloud Comput. 1(2), 20–24 (2014)
5. Carlini, E., Coppola, M., Dazzi, P., Ricci, L., Righetti, G.: Cloud federations in contrail. In: Alexander, M., et al. (eds.) Euro-Par 2011, Part I. LNCS, vol. 7155, pp. 159–168. Springer, Heidelberg (2012). doi:10.1007/978-3-642-29737-3_19
6. Buyya, R., Ranjan, R., Calheiros, R.N.: InterCloud: utility-oriented federation of cloud computing environments for scaling of application services. In: Hsu, C.-H., Yang, L.T., Park, J.H., Yeo, S.-S. (eds.) ICA3PP 2010. LNCS, vol. 6081, pp. 13–31. Springer, Heidelberg (2010). doi:10.1007/978-3-642-13119-6_2
7. NIST SP 500-292, Cloud Computing Reference Architecture. http://www.nist.gov/customcf/get_pdf.cfm?pub_id=909505.pdf
8. Demchenko, Y., Makkes, M.X., Strijkers, R., de Laat, C.: Intercloud architecture for interoperability and integration. In: 2012 IEEE 4th International Conference on Cloud Computing Technology and Science (CloudCom), Taipei, pp. 666–674 (2012)
9. Kamateri, E., Loutas, N., Zeginis, D., Ahtes, J., D'Andria, F., Bocconi, S., Gouvas, P., Ledakis, G., Ravagli, F., Lobunets, O., Tarabanis, K.A.: Cloud4SOA: A semantic-interoperability PaaS solution for multi-cloud platform management and portability. In: Lau, K.-K., Lamersdorf, W., Pimentel, E. (eds.) ESOCC 2013. LNCS, vol. 8135, pp. 64–78. Springer, Heidelberg (2013). doi:10.1007/978-3-642-40651-5_6
10. Zhang, L.J., Zhou, Q.: CCOA: cloud computing open architecture. In: IEEE International Conference on Web Services, ICWS 2009, Los Angeles, CA, pp. 607–616 (2009)
11. Zhang, H., Yang, X.: Cloud computing architecture based-on SOA. In: 2012 Fifth International Symposium on Computational Intelligence and Design (ISCID), Hangzhou, pp. 369–373 (2012)

Traffic Monitoring in Software Defined Networks Using Opendaylight Controller

Duc-Hung Luong[(✉)], Abdelkader Outtagarts, and Abdelkrim Hebbar

Nokia Bell Labs, Nozay, France
{duc_hung.luong,abdelkader.outtagart,
abdelkrim.hebbar}@nokia.com

Abstract. Software Defined Network (SDN) is an emerging approach in network technology appeared in recent years. In the SDN revolution, OpenFlow is defined as the first communication standard to separates the control plane and data plane this allows the control plane to be centralized by an OpenFlow controller. The rise of SDN and OpenFlow has changed the point of view of the conventional model for network devices. In this paper, we focus on QoS monitoring in SDN controller using OpenFlow protocol. We propose a monitoring method for collecting the statistic and calculating the throughput of link traffics. We also design a new forwarding algorithm for control plane that avoids bottleneck and provides load-balancing. To evaluate our algorithm in real time traffic, we setup a test-bed in Mininet where we design the applications using the devised algorithm in Opendaylight controller, a fully functional open-source project with a rich set of API.

1 Introduction

Nowadays, Software Defined Networking (SDN) [1] is becoming an emerging paradigm in network technology. SDN enables an easy way to control for the future innovation of network and cloud [2]. Enterprises deploy SDN because it promises speedy service provisioning flexibility and reduction of operating expenses. In SDN approach, the control planes are centralized to manage all the network devices. The controller also computes the routes and installs the corresponding rules to the remote switches. The installed rules are also changed to quickly adapt to workload or network infrastructure modifications. Hence, while the traffic engineering (TE) becomes more and more important in the network, the controller needs to continually monitor the network devices status and link states.

OpenFlow [3] is the dominant southbound protocol used for SDN context that connects the control plane and data plane. The OpenFlow controller can poll the switches by using statistic request messages and collect per-switch and per-flow statistics from these switches. However, the counter values from statistics messages are not sufficient for monitoring other QoS parameters such as network throughput and packet latency.

In this paper, we propose a solution to monitoring throughput performance, an important parameter of network QoS. Our contribution is to develop a traffic monitoring solution using OpenFlow features in a SDN network. Studying and choosing a

© Springer International Publishing AG 2016
S. Boumerdassi et al. (Eds.): MSPN 2016, LNCS 10026, pp. 38–48, 2016.
DOI: 10.1007/978-3-319-50463-6_4

suitable OpenFlow controller is an important task in our work. OpenDaylight [4], a Java-based open source controller has been chosen for implementing the monitoring application. This application focuses on the throughput monitoring and flow forwarding. The prototype demonstrates that this solution is effectively suitable in the SDN context.

The rest of this paper is organized as following. Section 2 provides a background and related works regarding the traffic monitoring solution in SDN networks. Section 3 presents our framework running on the controller OpenDaylight. We evaluate the performance of throughput monitoring through simulation using Mininet [5] in Sect. 4. Finally, we conclude the paper in Sect. 5 and point out some perspectives for our future works.

2 Background and Related Works

SDN is considered as the programmable networks with the ability to design and operate flexibly, thus, it has a lot of attentions for networking innovation. SDN's architecture has three-layer model shown in Fig. 1. The application layer includes all applications that are handled in the network and the infrastructure layer refers to the data plane, which has been decoupled from the controller. The controller can be considered as the "brain" of the networks. Its role is the control point to relay information to the hardware below via *Southbound APIs* and to the applications above via *Northbound APIs*.

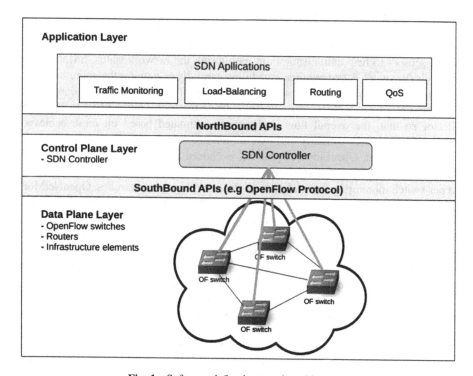

Fig. 1. Software defined network architecture.

Currently, traffic engineering (TE) is an important paradigm that provides the mechanism to manage, analyze and predict network behavior. As a result, network administrator can optimize performance and improve reliability of the network. Learning from the past, there are some existed solutions of traffic engineering for conventional network such as Asynchronous Transfer Mode (ATM) [6] network or Multi-protocol Label Switching (MPLS) [7]. However, the emergence of SDN paradigm that centralizes the control plane of distributed network devices requires new intelligent and adaptive TE mechanism. In [8], the authors consider that the scope of traffic engineering consists in four parts: monitoring for flow management, fault tolerance of network, topology update and traffic analysis. In this paper, traffic monitoring for flow management approach will be considered and focused.

In traffic monitoring paradigm, the monitoring application requires accurate and timely link statistics and device states. The traditional traffic monitoring using two approaches for measurement methods: *active measurement* and passive *measurement*. The *active measurement* method injects additional packets in to networks and monitors its behaviors. The simple example is the use of *"ping"* to determine the end-to-end connection status, compute the packets loss and discover the topology of networks. However, this measurement method produces interferences in to networks as well as being a cause of overhead. As opposed to *active measurement, passive measurement* is another approach for network monitoring while it does not inject traffic in to networks. This method measures the network traffic by observation and without generated overhead thus it does not influence network performances. However, passive method does not achieve results with good accuracy and is not always feasible for all networks.

Currently, there are some measurement protocols for the conventional network such as SNMP (RFC 1157) [9], NetFlow [10] from Cisco, sFlow [11] from InMon... Simple Network Management Protocol (SNMP) is an application layer protocol that uses passive sensors to help administrators in monitoring the network status. SMNP regularly polls the switches and requires scheduling carefully to monitor the entire network. However, SMNP is not suitable for flow-based monitoring and will not be considered in the present paper. NetFlow periodically collects the traffic information to NetFlow collector so that, the overall flow statistics are estimated based on these achieved samples. SFlow works similarly and uses time-based to collect packet sampling.

In SDN context, OpenFlow [3] protocol is chosen as the communication standard between the controller and network devices. OpenFlow enables controller to per-flow and per-switch monitoring that picked up in the several recent researches. OpenNetMon [12] is an active measurement tool designed as a Python-based module for POX controller [13]. This application monitors per-flow quality of service (QoS) metrics by adaptively polling switches at adaptive rates. OpenNetMon uses an adaptive method for changing the polling interval of statistic requests. Presented as a passive method, FlowSense [14] uses the controller messages such as *PacketIn* and *FlowRemoved* to manage the network and predict per flow link utilization. This method enables monitoring the SDN networks with zero measurement cost. Instead of actively querying the switches, the controller only uses the information from *PacketIn* messages and *FlowRemoved* messages. The traffic parse module in the controller captures flow traffic hence controller estimates the traffic utilization. FlowSense proposes a zero cost approach to monitor networks; however the estimation of FlowSense obtains inaccurate results due to long idle-time out before receiving *FlowRemoved* message. Payless [15] is another query-based monitoring framework for flow statistic collection of different levels. Payless monitoring provides a set of well-defined API which is very useful for different network applications to monitor and collect data based on it. OpenTM [16] is a monitoring scheme that estimates the traffic matrix (TM) of OpenFlow networks. This module is an application implemented for NOX, the first open-source OpenFlow controller. OpenSample [17] proposes the mechanism to monitor in real-time, low latency and flexible. It takes the advantage of sFlow to monitor high-speed networks.

In order to satisfy our context, we also study for choosing adaptable SDN controller. There are variety of OpenFlow controllers, with different languages and supported environments [18, 19]. NOX [20] is the first OpenFlow controller, but it is not the heavily implemented or used. NOX is a multi-threaded and C++ based program. Although NOX supports graphical interface and visualization tool, however its weakness is lack of documentation and performance. POX [13] is a Python-based OpenFlow controller inherited from NOX controller. In comparing to NOX, POX is more familiar to implement and use because of providing a web based interface. Beacon [21] is another Java-based OpenFlow controller that supports both event-driven and multi-thread. Same as Beacon, Floodlight [22] is Java-based framework under Apache licensed. Its core architecture is modular and provides many components, including topology management, device management, path computation and web access. Ryu [23] is a component-based frame work which has a set of predefined components. Ryu is implemented in Python and is supported by NTT labs. The components of Ryu can be modified, extended and customized for specific application. More recently, ONOS [24] offer a distributed SDN controller with similar features. However, ONOS focuses on specific tasks of service providers and will be not considered. In this paper, we investigate Opendaylight [4] for implementing our monitoring algorithm. Opendaylight is a Java-based open-source controller supported by many network vendors. Furthermore, Opendaylight has highly modular architecture with a rich set of APIs and features for SDN network.

3 System Design

3.1 Methodology

In an OpenFlow controller, the controller regularly queries *StatisticsRequest* message to retrieve the statistics of switches, ports, flows and flow tables of network switches. In the present work, we focus on the statistics of switch ports and flows via *StatisticReply* messages.

However, as specified in OpenFlow standard, we cannot retrieve directly the values of throughput of ports and flows. The OpenFlow controller only polls the statistics of switches by the number of transmitted packets or transmitted bytes. Hence, we propose a solution which regularly retrieves the traffic amount and consider that the throughput is equal to the average transmission rate by the time unit. In this work, our module receives the amount of S(bytes) in duration of time t(s) hence, the throughput is calculated as:

$$R = \frac{S}{t} (byte/s) \tag{1}$$

Polling of flows and ports is made regularly by the controller. The value of the interval duration is the same for each path in the network. In Opennetmon [12], the authors use the same solution in the monitoring module of POX controller. However, they consider the random or round robin policies that are not suitable for large scale network. In the present work, we only use the round robin policy to facilitate the solution because of the small network environment.

3.2   Architecture Design

In this section, we describe two modules for Opendaylight controller for throughput monitoring and packet forwarding in OpenFlow-based network.

Throughput monitor. As presented in the previous section, OpenFlow is a vendor-neutral standard communication that defines the interaction between controllers and switches. OpenFlow protocol-plugin is the southbound module that connects directly with the switches (real or virtual switch). In this work, the monitoring module regularly requests the OpenFlow protocol plugin to send *StatisticsRequest* messages to the switches. There are some types of statistics message such as *vendor, flows, table, port* in which we are interested to request the port state statistics. The switch that receives this request responds by a *StatisticsReply* message. More specifically in Opendaylight, OpenFlow protocol-plugin module implements services to capture and parse this message from switches. In our case, the information on switch ports and flows carried by the statistic message are collected and processed by the statistics module. We mainly focus on the number of packets and bytes passed by all ports to calculate the throughput. By storing the packets count and bytes from previous port state, the delta of these counters is obtained to determine the current transmission rate for each link. These values of throughput statistics are locally stored in the OpenFlow plug-in module. Another database is created in the controller to store all the statistics of port states and flow states. When OpenFlow protocol plug-in detects a change of a statistics counter, it automatically notifies the upper layer to update the database. Inside of the controller core, the AD-SAL module provides another service to read the statistics of all components in the network. Additionally, we modified the *StatisticsManager* module to retrieve the throughput data. This module also provides the APIs to the monitoring application. The user also retrieves the data by using REST - client via northbound interfaces (Fig. 2).

Packet Forwarding. Actually, when a conventional switch receives a packet, it learns the mapping between MAC address and port to match. If the switch already knows the specific destination, the packet is sent directly to the correct target switch port. Otherwise, the packet is flooded out all ports like in a hub. This functionality is called Layer 2 learning switch. In case of a flow-based switch managed by a controller, when a switch receives a new packet that is not matched with any installed flow table, it encapsulates the payload into a *PacketIn* message and sends this message to the

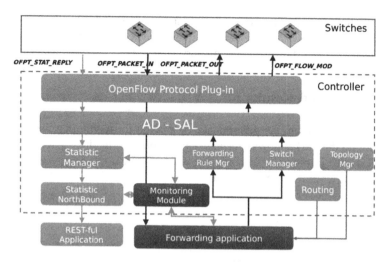

Fig. 2. Diagram of monitoring module and forwarding in Opendaylight

controller. The controller responds for installing a path using one or more *FlowMod* message and resends another *PacketOut* to the destination switch. In our application, we consider the routing in Layer 3 to find the shortest path. It is also looking for a new path while link utilization value of the shortest path is high. The decision to modify the packet path is based on the throughput result computed by the monitoring module. This method allows avoiding the collision and packet loss because the full link capacity is used at main path. It also provides the network load balancing because based on changing the path to the destination.

The algorithm used for monitoring and forwarding works as following. Firstly, the controller creates a list of rules that will be processed to install the flow. When the controller receives a *PacketIn* message, the application extracts the IP address of the destination host. Then, it looks up the destination host (*dest_host*) which corresponds to this IP address from the IP host table. The current switch (*curr_switch*) that sent *PacketIn* and the destination switch (*dest_switch*) that connects to this host is also tracked. By using Dijkstra algorithm [25], we can find the shortest path between *curr_switch* and *dest_switch*. This solution determines a unique path to destination and loop-avoiding in the network. From this path, the *next_switch* is defined as the next hop that connects directly with the *curr_switch*. The new flow is created with the IP address of the destination and the action is set to the egress port of *curr_switch*. We create a new rule that indicates that the flow will be processed by *curr_switch*. This rule is also added to the list of rules in order to the controller fabricates a new *FlowMod* message for *curr_switch*.

The results from monitoring module allow us to calculate the '*curr_switch – next_switch*' link throughput value and to compare this value to a fixed threshold. In case the link throughput exceeds the threshold, the link between *curr_switch* and *next_switch* is temporary disabled and another shortest path is computed. From this new path, we find the *new_next_switch* and the link respectively. We continuously compute the link throughput of the new link and compare it to the old link between *curr_switch* and *next_switch*. If the capacity rate of the new link is higher than the older one, we install the new flow on the temporary path.

4   Evaluation and Results

In this section, we present the performance of a demo for monitoring the throughput and forwarding the packet flows. For this application, two modules have been developed as described in previous sections. We will consider a topology with 6 switches and two scenarios for the demo (Fig. 3).

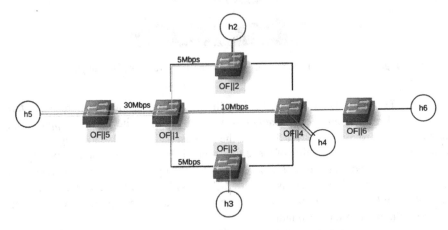

Fig. 3. Forwarding scenario for monitoring module

4.1   Evaluation

Monitoring module. Firstly, we consider the monitoring scenario while we try to capture the real-time traffic of link *OF1 – OF5*. In this scenario, the network topology is shown in the Fig. 3. We set the bandwidth of links between OF1 – OF2, OF1 – OF3, OF1 – OF4, OF1 – OF5 at 5 Mbps, 10 Mbps and 30 Mbps respectively. We use *Iperf* [26] packet generator to send and receive TCP packets between a pair of hosts. By running in client-server mode, *Iperf* will send TCP packets traffic to specific hosts at specific bandwidths. For example, H5 host runs *Iperf* at server mode and other hosts run *Iperf* in client mode. Traffic is generated from H2, H3, H4, to H5 with transmission rates of traffic equal to the max capacity value allowed on each link.

Figure 4 shows the results of the simulation during 1000 s. We firstly send 5 Mbps of traffic from H2 to H5 at time t = 0 s using *Iperf*. After 300 s, we continuously transmit the traffic from H3 to H5. This traffic has a transmission rate equal to the max traffic between H3–H5. Hence, from t = 300 s, H5 receives the traffic at 10 Mbps that

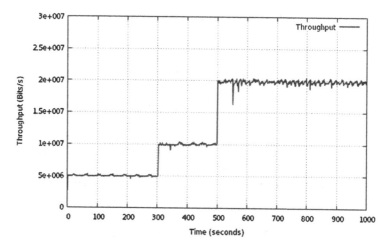

Fig. 4. Results of throughput monitoring

equals to the cumulative traffic sent from H2 and H3. Similarly, at t = 500 s we generate the traffic from H4 to H5 and the total traffic of link H1 – H5 consequently increases until 20 Mbps.

Forwarding scenario. In this scenario, the link bandwidths are set equally to 10 Mbps for every link in the network and the demonstration considers the connection path between H5 and H6 to verify the algorithm (Fig. 5).

Firstly, we consider the normal scenario to obtain the default behavior of network. We send the ICMP packet via the *ping* command in Mininet to confirm the paths. The shortest path found from H5 to H6, using Dijkstra [25] routing algorithm, is via OF1, OF4 and OF6. The flows installed in the switches are listed in Table 1. The Web GUI of Opendaylight can be used to check that the flows have been successfully installed.

Now we consider the load balancing scenario. We clear all the existed flows in the network via RestClient application and northbound API (Fig. 5). We also set the threshold value for changing the path at 80%. Then we generate some traffic between OF1 and OF4 such as the threshold is exceeded.

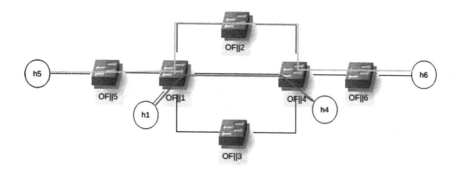

Fig. 5. Forwarding scenario for load balancing

Table 1. Flows created in normal scenario

Switch	IP of destination	Action
OF5	10.0.0.6	Output = Port 2
OF1	10.0.0.6	Output = Port 4
OF4	10.0.0.6	Output = Port 5
OF6	10.0.0.6	Output = Port 6

When the traffic between these switches passes the threshold value, the link between OF1 and OF4 is temporary set unavailable and the controller looks up other paths for load balancing. The result can be verified when we ping again *h5 ping h6*, the path between H5 and H6 is changed by the load balancer. As a result, we obtain automatically the new shortest path H5 – OF5 – OF1 – OF2 – OF4 – OF6 – H6 (Fig. 5) and the flow table of these switches has been also changed to complete the path installation (Figs. 6 and 7).

Node	In Port	DL Src	DL Dst	DL Type	DL Vlan	Vlan PCP	NW Src	NW Dst	ToS Bits	NW Proto	TP Src	TP Dst	Actions	Byte Count	Packet Count	Duration Seconds	Idle Timeout	Priority
OF|00:00:00:00:00:00:00:01	*	*	*	IPv4	*	*	*	10.0.0.1	*	*	*	*	SET_DL_DST = fe:e7:91:56:a1:7f OUTPUT = OF|1	294	3	44	0	1
OF|00:00:00:00:00:00:00:01	*	*	*	IPv4	*	*	*	10.0.0.6	*	*	*	*	OUTPUT = OF|4	98	1	8	0	1
OF|00:00:00:00:00:00:00:01	*	*	*	IPv4	*	*	*	10.0.0.4	*	*	*	*	OUTPUT = OF|4	98	1	44	0	1

Fig. 6. Flow table of switch OF1 in default scenario

Node	In Port	DL Src	DL Dst	DL Type	DL Vlan	Vlan PCP	NW Src	NW Dst	ToS Bits	NW Proto	TP Src	TP Dst	Actions	Byte Count	Packet Count	Duration Seconds	Idle Timeout	Priority
OF|00:00:00:00:00:00:00:01	*	*	*	IPv4	*	*	*	10.0.0.1	*	*	*	*	SET_DL_DST = 02:12:84:19:2c:4c OUTPUT = OF|1	32813016	7592	155	0	1
OF|00:00:00:00:00:00:00:01	*	*	*	IPv4	*	*	*	10.0.0.4	*	*	*	*	OUTPUT = OF|4	501312	7556	151	0	1
OF|00:00:00:00:00:00:00:01	*	*	*	IPv4	*	*	*	10.0.0.6	*	*	*	*	OUTPUT = OF|2	196	2	155	0	1

Fig. 7. Flow table of switch OF1 in load balancing scenario

4.2 Discussion

In this implementation, we design the traffic monitoring application using OpenDaylight controller. This application automatically monitors the bandwidth utilization of the links in the network and reactively chooses the best alternative path to destination. The forwarding component tracks the host location to install per-flow forwarding rules on the switches. This module also learns from the real-time traffic obtained by monitoring module to find other paths. In comparing with existing forwarding algorithms, we brought some new specific differences detailed below.

As the switches process too many flows, a bottle neck may occur. When the amount of data passing through one link in a short time is too large, the latency and packet loss will be clearly considered. In case of bottle neck, we chose a temporary path for load balancing and reduction of packet lost rate. We also analyzed the traffic based-on bandwidth utilization rate. This is an important indication to evaluate the Quality of Service (QoS) of a network. The installation of per-flow rule is done by all switches between *curr_switch* and *dest_switch* and iterates until all switches are configured. This method allows avoiding the repeated request of switches to the controller. We find a temporary path at a given moment but without changing the topology to adapt the dynamic of network. The link with high utilization will be considered at other moment to look up the best path. The forwarding module considers the switch acting as Layer 3 switch, that uses the routing module to find the shortest path based on Dijkstra algorithm. This is a simple case of routing algorithm but this mechanism allows choosing only a single path to host switch and avoiding loop in topology.

5   Conclusion and Perspectives

As the Software Defined Networking becomes an important part of the future network architecture, we must consider traffic monitoring as one of the important goals to adapt quickly to the network traffic changing. In this paper, we offer a solution for traffic monitoring using OpenFlow in Software Defined Networking. While there are many management protocols in SDN such as SMNP, NetFlow, sFlow…, OpenFlow is considered as a suitable standard to monitor the traffic in SDN networks. We also study about OpenFlow framework and we choose OpenDaylight as the best OpenFlow controller to deploy our implementation. The application consists of two main components: monitoring module and forwarding module. In the monitoring module, we regularly query the switch and calculate the flow bandwidth. The forwarding module tries to use the results of monitoring module to install per-flow forwarding rules. The shortest path is found by using Dijkstra algorithm to obtain the best path and avoid loops in the topology. In the last section, we have shown experimental results using our test bed.

The application of traffic monitoring in SDN networks presents many challenges. In this paper, we only use throughput as the QoS parameter for monitoring the traffic. For the future works, we will investigate other parameters such as latency and packet loss. The forwarding module in this paper changes the path at the initialization of flow. We presented a simple scenario for changing the path in the topology. The more complex scenarios such as automatically deleting a flow, deleting a link are also in the perspectives for the future of our works.

References

1. Open network foundation: software-defined networking: the new norm for networks. https://www.opennetworking.org/
2. Weldon, M.K.: The Future X Network: A Bell Labs Perspective. Nokia BellLabs, Murray Hill (2016)

3. McKeown, N., Anderson, T., Balakrishnan, H., Parulkar, G., Peterson, L., Rexford, J., Shenker, S., Turner, J.: OpenFlow: enabling innovation in campus networks. SIGCOMM Comput. Commun. Rev. **38**(2), 6974 (2008)
4. OpenDaylight controller. https://www.opendaylight.org/
5. Mininet. https://www.mininet.org/
6. de Prycker, M.: Asynchronous Transfer Mode: Solution for BroadBand ISDN, Ellis Horwood, Ltd., (1993)
7. Rosen, E., Viswanathan, A., Callon, R.: Multiprotocol Lable Switching Architecture. Internet draft, draft-ietf-mpls-arch-01.txt (1998)
8. Akyildiz, I.F., Lee, A., Wang, P., Luo, M., Chou, W.: A roadmap for traffic engineering in SDN - OpenFlow networks. Comput. Netw. **71**, 1–30 (2014)
9. Simple Network Management Protocol. https://www.ietf.org/rfc/rfc1157.txt
10. NFC 3954: Cisco system NetFlow Services Export Version 9. http://tools.ietf.org/html/rfc3954.html
11. Phaal, P., Lavine, M.: sFlow Version 5. http://www.sfow.org/
12. van Adrichem, N.L.M., Doerr, C., Kuipers, F.A.: Opennetmon: network monitoring in openflow software-defined networks. In: Network Operations and Management Symposium (NOMS) (2014)
13. POX controller. https://www.noxrepo.org/pox/
14. Yu, C., Lumezanu, C., Zhang, Y., Singh, V., Jiang, G., Madhyastha, H.V.: FlowSense: monitoring network utilization with zero measurement cost. In: Roughan, M., Chang, R. (eds.) PAM 2013. LNCS, vol. 7799, pp. 31–41. Springer, Heidelberg (2013)
15. Chowdhury, S.R., Bari, M.F., Ahmed, R., Boutaba, R.: PayLess: a low cost network monitoring framework for software defined networks. In: Proceedings of the 14th IEEE/IFIP Network Operations and Management Symposium, NOMS 2014 (2014)
16. Tootoonchian, A., Ghobadi, M., Ganjali, Y.: OpenTM: traffic matrix estimator for OpenFlow networks. In: Krishnamurthy, A., Plattner, B. (eds.) PAM 2010. LNCS, vol. 6032, pp. 201–210. Springer, Heidelberg (2010)
17. Suh, J., Kwon, T., Dixon, C., Rozner, E., Felter, W., Carter, J.: OpenSample: a low latency, sampling-based measurement platform for SDN, IBM report (2014)
18. Tootoonchian, A., Gorbunov, S., Ganjali, Y., Casado, M., Sherwood, R.: On controller performance in software-defined networks. In: Hot-ICE12 Proceedings of the 2nd USENIX Conference on Hot Topics in Management of Internet, Cloud, and Enterprise Networks and Services (2012)
19. Khondoker, R., Zaalouk, A., Marx, R., Bayarou, K.: Feature-base comparison and selection of Software Defined Networking (SDN) controllers. In: World Congress on Computer Applications and Information Systtems (WCCAIS) (2014)
20. NOX controller. https://www.noxrepo.org/
21. Beacon controller. https://www.standford.edu/display/Beacon/Home/
22. FloodLight controller. https://www.projectfloodlight.org/floodlight/
23. Ryu controller. https://www.github.io/ryu
24. Open Network Operation System (ONOS). http://www.onosproject.org
25. Dijkstra, E.W.: A note on two problems in connexion with graphs. Numer. Math. **1**, 269–271 (1959)
26. Iperf, The network bandwidth measurement tool. http://iperf.fr/

Virtualization Techniques:
Challenges and Opportunities

Lyes Bouali[1,2(✉)], Emad Abd-Elrahman[3], Hossam Afifi[3],
Samia Bouzefrane[2], and Mehammed Daoui[1]

[1] LARI Lab, UMMTO, Tizi Ouzou, Algeria
boual_ly@auditeur.cnam.fr,
mdaouidz@yahoo.fr
[2] CEDRIC Lab, CNAM, Paris, France
samia.bouzefrane@lecnam.net
[3] RST Department, Institute Mines-Telecom, Saclay, France
{emad.abd_elrahman,hossam.afifi}@telecom-sudparis.eu

Abstract. The gap between the deployment costs for the new services, apps and hosting machines with the expected revenue forces the operators to think in the virtualization concept although the heterogeneity in virtualization techniques solutions. While this virtualization architecture came with different challenges like instances isolation in the hardware or software layers, it anticipated many opportunities for fast deployment and resources management. This position paper lists the main challenges starting from tenants' isolation and hardware trusting for virtualization layer till the software and network mapping issues for virtual instances communication under the softwarization techniques. Moreover, it sums up the directions for this research point in terms of the vision of software and hardware.

Keywords: Virtualization · VM · Containers · TPM · SDN · NFV

1 Introduction

Most IT systems consists of different technologies for shared virtualized infrastructures that include hybrid cloud and data centers that rely on virtual machines based hypervisors and containers based solutions to achieve certain goals such as load balancing and fault tolerance by taking advantage of VM migration and also reduce the consumption of energy by the consolidation of servers [1].

Since the first introduction in the mid of 1960s by IBM and the increasing of cloud technology, virtualization has involved extremely the paradigm of computing technology. In fact, the virtualization concept evolved from software-based technique to, more recently, hardware-based solution by virtualizing memory, processor and devices more efficiently, targeting currently other fields such as the mobile and embedded platforms [2, 3].

© Springer International Publishing AG 2016
S. Boumerdassi et al. (Eds.): MSPN 2016, LNCS 10026, pp. 49–62, 2016.
DOI: 10.1007/978-3-319-50463-6_5

In addition, with the appearance of the new virtualization technology such as isolation container, applications may be run in isolated environments with a minimum overhead unlike the traditional virtualization solutions. Due to the open and the interoperation of current computing systems, the trusted computing has been widely designed to secure the integrity of the systems from software attacks and a few hardware attacks. Trusted computing, which is based on hardware component such as Trusted Platform Module (TPM), can be also recognized as a set of technologies that provide a primitive security mechanism to protect computing infrastructure. In general, TPM [4] is a hardware platform dedicated to PCs, servers, personal digital assistant (PDA), printer, or mobile phone to enhance security in an ordinary, non-secure computing platform and convert them into trust environment. Each platform contains only one physical TPM to be a trusted platform. TPM implements mechanisms and protocols to ensure that a platform has loaded its software properly. This has been named as remote attestation. TPM is independent from the host operating system (OS); it stores secret keys to encrypt data files/messages, to sign data, etc.

By taking advantages of virtualization technology benefits, physical TPM delivers its functionality to the multi guest operating system in the abstract way. As a result, TPM virtualization provides security architecture in the virtualizable platforms.

Virtualization in networking is not a new concept. In fact, VLANs (Virtual local area networks) can be built with isolation on a physical LAN shared by multiple entities of a company. Similarly, virtual private networks (VPNs) use generally Internet to connect to a private network, hence offering the same level of security as if the companies and employees are accessing directly private networks. In addition, some research works have been carried out to introduce virtualization in wireless networks [5]. However, the renewed interest in network virtualization has been pushed by the cloud computing concept. In fact, in this new context, many new notions related to virtualization have emerged like the virtualization of NICs, switches, or network functions [6] to refer to the NFV (Network Function Virtualization) standard. Hence, the ETSI standard proposes a specification to define a reference Network Function Virtualization infrastructure. Software-Defined Networking (SDN) is another concept that also helps in network virtualization.

In this article, our aim is:

- Firstly, to highlight the different virtualization techniques used by the virtual machine manager (VMM) and the alternative solutions that may offer better performances,
- Secondly, to explore the virtualization in networking by focusing on new concepts like NFV and SDN while demonstrating that a trusted hardware can improve VMs' security, and
- Finally, to list the main open issues that are generated by the softwarization techniques.

The rest of this paper is constructed by the following sections. We first present the general principles of virtualization in Sect. 2. Section 3 describes the concepts behind network softwarization like SDN or NFV. Section 4 shows that virtualization concept can be extended to the security aspect by virtualizing trusted hardware to offer secure virtualized architecture. Before concluding in Sect. 6, Sect. 5 points on virtualization challenges in terms of heterogeneity, isolation, security and performance.

2 Virtualization

Nowadays, virtualization technology is widely used to share the capabilities of physical computers by splitting the resources among operating systems. In 1964, the concept of virtual machines (VMs) is used in the IBM's project called CP/CMS system. The Control Program (CP) acted to split the physical machine to several copies (VMs), each copy with her own OS CMS. In 1974, Goldberg and Popek published a paper [7] which introduces a set of sufficient conditions for computer architecture to efficiently support system virtualization. In 1999, the company VMware presented its product VMware Virtual Platform for the x86-32 architecture [8]. In 2006, Intel and AMD proposed extensions to support virtualization. In the next subsections, we will recall the basic principles of virtualization.

2.1 Virtualization Techniques

The virtualization layer called hypervisor or *Virtual Machine Monitor* (VMM) manages the hardware resources between the different VMs. The guest OS is the OS that is running within a VM. Currently, there are several virtualization techniques that we summarize in the following.

Full Virtualization
In full virtualization, the source code of the guest OS is not modified. So, when executed, the guest OS is not aware to be virtualized. Binary translation (BT) is used to emulate a small set of instructions, i.e., privileged instructions are trapped and sensitive but non-privileged instructions are detected then translated to show a fake value. The rest of the instructions are directly executed by the host CPU. There are two types of hypervisors that handle the full virtualization:

Type 1 corresponds to a native or standalone hypervisor that runs directly on the bare hardware. Type 2 is a hypervisor that runs on top of the host OS (like an ordinary application).

The full virtualization benefit is that guest OSs may run on a VMM while keeping unmodified. But, the drawback is that the VMM must use special tricks to virtualize the hardware for each VM, which generates an additional overhead when accessing the hardware.

Para Virtualization
It refers to the collaboration between the guest OS and the hypervisor to improve the performance. This collaboration involves modifying the source code of guest OS to call directly, using hyper calls, the hypervisor to execute privileged instructions. So the guest OS is aware to be virtualized. The drawback of this technique is the poor compatibility and portability of OSs.

Hardware Assisted Virtualization

To simplify the virtualization techniques, hardware vendors such as Intel and AMD introduced new extensions to support virtualization. Hence, two mode features for CPU execution are introduced in Intel Virtualization Technology (VT-x) and AMD's AMD-V so that the VMM runs in a root mode below ring 0 (i.e., ring −1) while the guest OS runs in non-root mode (ring 0). The state of the guest OS is stored in Virtual Machine Control Structures (for Intel VT-x) or Virtual Machine Control Blocks (for AMD-V). The advantage of the hardware assisted virtualization is the reduction of the overhead caused by the trap-and-emulate model [9, 10].

2.2 Isolation by Containers

As stated in the preceding section, hypervisor-based virtualization is the adopted approach in the virtualization domain, and many known solutions are based on that model such as VMware solutions, Xen, VirtualBox, etc. This approach is based on virtual machines, where each VM contains a full operating system with all of its layers. The hypervisor scheduling operates at two levels as explained in the following. The first scheduling occurs at the hypervisor level where the hypervisor schedules VMs, and the second one at the virtual machine level where the guest operating system schedules processes. Taking that into account, it is obvious that this approach creates a significant overhead. Another virtualization approach, based on containers, has been proposed to reduce significantly this overhead, and thus providing better performances, especially in terms of elasticity and density within a lean data center.

In the container based virtualization, scheduling occurs only at one level. The containers contain processes, while the kernel and the libraries are shared between the containers. Hence, the scheduler is used only at process level to schedule processes. The behavior is similar to a classical operating system; the only difference is that, in container-based virtualization, processes are affected to a container, and between each container, an isolation is provided by the operating system.

We can define a container as a confined environment under the global environment. The Linux operating system provides some mechanisms to create these confined environment or container, and somehow forces each container to follow a predefined policy.

The isolation can occur at two levels: isolation between processes using *namespaces* mechanism [11], and isolation between processes and the hardware by relying on the *cgroup* (abbreviated from control groups) mechanism [12]. The combination of the *namespaces* and *cgroup* Linux functionalities allow the design of containers which are isolated from both the process and the hardware perspectives. Even if these functionalities are developed by different teams, the two Linux solutions are orthogonal, and most of the container-based virtualization solutions like Lxc and Docker rely on them.

3 Network Virtualization

Network virtualization can be defined as the building of multiple logical networks over a physical network infrastructure. The output is a group of nodes (either virtual machines or containers) that connected together over the top of hypervisors or Docker containers using virtualized infrastructure. The links between the created instances can be assured either using internal direct communication or through a specific tunnel.

3.1 Network Virtualization Topologies

Logical Topologies

The emulation of logical network structure over physical one can be manipulated either using layer 2 or layer 3 techniques for network based architectures.

- For layer 2 techniques, Virtual Local Area Network (VLAN) is a standard isolation technique known in the network many years ago. VLAN is a logical grouping of several stations in the same broadcast domain (Broadcast) regardless of their geographical location. The administration and management of VLANs are easier than managing classical LANs, knowing that the organization is based on a logical dependency of the director, not the physical location of stations.
- For layer 3 techniques, the Virtual Private Network (VPN) is a good emulation and it can also be emulated in Layer 2. VPN can be defined as an overlay network built over public networks like Internet. It is generally used to connect one or more sites using a secure tunnel either generated using layer 2 tunnel protocols like L2TP, VPLS for MPLS technology or layer 3 tunnel protocols like IPSec.
- Another logical topology structure based application is the point to point architectures P2P. It is an overlay network built on top of one or more physical networks. The majority of overlay networks are implemented at the application level. Different implementations exist at other levels, to ensure some type of network functions: like multicast routing, Service Quality.

Virtual Topologies

The principle in virtualization is to provide the customer a point to point or multipoint network overlay on the physical infrastructure. However the user has no real control over the implementation of this network. With the recent emerging concepts of communication, the using of traditional VLANs or VPNs is not coping with the data centers. Indeed, the scalability problems are prohibitive in data centers due to the increase in the number of virtual machines:

- Limitations of spanning-tree protocol (number of VLANs, convergence time of the spanning-tree),
- Difficulty in communicating or migrate VMs from different VLANs on different physical machines.

These issues also related to the creation of offers customers including network connectivity and development of multi-tenant offerings (or multiple customers share the infrastructure) have stimulated the development of network virtualization technologies to create topologies level 2 large scale.

Different approaches exist based mainly on the construction of a logical network level 2 from an address translation mechanism or encapsulation on an L2 or L3 layer. The implementations of such mechanisms involve network management techniques to determine the logical addresses and match them to the MAC addresses of virtual machines.

- Network Virtualization using Generic Routing Encapsulation (NVGRE) is the creation of a Layer 2 network to overlay on a network level 3. The Ethernet frames are transported in a tunnel Generic Routing Encapsulation (GRE) through an IP network. NVGRE connects machines belonging to different IP networks by making them belong to the same level of logical network 2. The GRE header scalability is enough to reproduce the proper diffusion mechanisms to a Layer 2 network, using IP multicast addresses (in terms of its number of bits mapping). Therefore, scaled problems generated by a large number of virtual machines connected via Ethernet are partially settled through the mechanisms for IP addressing. Figure 1 shows an example of GRE encapsulation for two VMs communication (10.1.1.2 to 10.1.1.1) re-encapsulated using network 192.168.0.0 for outer headers.

```
14 5.999550    10.1.1.2              10.1.1.1             ICMP                136 Echo (ping) reply
▷ Frame 1: 136 bytes on wire (1088 bits), 136 bytes captured (1088 bits) on interface 0
▷ Ethernet II, Src: Dell_7a:d3:21 (00:21:9b:7a:d3:21), Dst: Dell_04:e2:74 (00:22:19:04:e2:74)
▷ Internet Protocol Version 4, Src: 192.168.0.155, Dst: 192.168.0.152
▷ Generic Routing Encapsulation (Transparent Ethernet bridging)
▷ Ethernet II, Src: b2:00:13:f6:59:4f (b2:00:13:f6:59:4f), Dst: aa:d8:d3:a7:db:4e (aa:d8:d3:a7:db:4e)
▷ Internet Protocol Version 4, Src: 10.1.1.1, Dst: 10.1.1.2
▷ Internet Control Message Protocol
```

Fig. 1. GRE encapsulation through Wireshark capture.

Virtual Extensible LAN (VXLAN) is a very close NVGRE solution and is well supported by Microsoft. It consists of encapsulating the Ethernet frames in UDP (reserved port 4789 as shown in Fig. 2). The tunnel terminates at the hypervisor and at the host server. Besides Microsoft, including VXLAN is supported by Cisco and VMWare. It provides the required scalability in terms of VLANs extension by increasing the number of VLANs.

```
99 10.024011   192.168.1.30          192.168.1.3          ICMP                148 Echo (ping) reply
▷ Frame 56: 148 bytes on wire (1184 bits), 148 bytes captured (1184 bits) on interface 0
▷ Ethernet II, Src: Dell_04:e2:74 (00:22:19:04:e2:74), Dst: CiscoInc_4a:cc:01 (64:9e:f3:4a:cc:01)
▷ Internet Protocol Version 4, Src: 157.159.124.97, Dst: 157.159.103.93
▷ User Datagram Protocol, Src Port: 53073 (53073), Dst Port: 4789 (4789)
▷ Virtual eXtensible Local Area Network
▷ Ethernet II, Src: 3comCorp_b7:24:d5 (00:50:04:b7:24:d5), Dst: Broadcom_60:07:39 (00:10:18:60:07:39)
▷ Internet Protocol Version 4, Src: 192.168.1.3, Dst: 192.168.1.30
▷ Internet Control Message Protocol
```

Fig. 2. VXLAN encapsulation through Wireshark capture.

- Stateless Tunnelling Transport (STT) is an alternative of the same type of NVGRE and VXLAN. It is based on TCP headers to encapsulate Ethernet frames. TCP headers are used to take advantage of specific TCP segmentation mechanisms implemented in hardware network interfaces machines but the protocol is actually stateless. STT is supported by Nicira and is not a solution adopted to be standardized.

- Transparent Interconnection of Lots of Links (TRILL [13]) is a network protocol virtualization in data centres being standardized at the IETF. It uses the level 3 routing techniques to implement routing mechanisms to layer-2. The routing protocol used is the ISIS protocol. TRILL is based on the implementation of R-bridges that are network devices that support the TRILL protocol. TRILL can be compared to previous protocols in the sense that it performs an encapsulation of Ethernet frames but this time in another Ethernet frame layer-2 e (MAC in MAC encapsulation). It also adds a virtual network identifier of 24 bit that increases the number of logic networks of the standard VLAN. The main advantage of TRILL is its layer-2 of network transmission. It does not require the implementation of an IP subnet with its own control plan and that requires configuration and an addressing ad-hoc plane. The MAC in MAC encapsulation does not imply any change IP addresses in case of machine migration. The TRILL solution is also compatible with standard layer-2 facilities and is transparent to the IP packet layer-3.

3.2 NFV and SDN

The current emerging technologies for emulating the virtualization of network function can be detailed as follows:

NFV

NFV (Network Function virtualization) was introduced to face a number of challenges in TSPs (Telecommunication Service Providers) such as the aim to build more dynamic, flexible and service-aware networks and to reduce the CAPEX/OPEX (CAPital EXpenditure/OPeration EXpenditure) [14]. By using the virtualization technologies, NFV decouple the NFs (Network Functions) from the devices on which they are traditionally run. This approach allows the consolidation of several NFs on standard server, switches and storage devices which may be located on data centers or at End User Premises. In this way, a service would be broken down into a set of VNFs (Virtualized Network Functions) that could be run as software on standards servers. Therefore, these VNFs could be relocated in other networks [15–17].

SDN

SDN (Software-Defined Networking) aims to reduce the CAPEX/OPEX costs and to make networks more flexible by providing an open and user-controlled use of the network's forwarding devices (routers and switches) [18]. The main concept of SDN is the split of the control plane (Network Operating System or controller) from the data plane (forwarding plane) in the forwarding devices. In that way, the control plane, which becomes a software executed on standard servers, sends instructions to be performed by the data plane which is composed of SDN-enabled routers and switches. This approach provides a global view of the network and simplifies its management. In addition, SDN provides programmability to share the network infrastructure between several virtual networks with different policies and allow APIs standardization [6].

3.3   Relation Between SDN and NFV

Actually, SDN and NFV aim at the same objectives, they are highly complementary but they are not dependent of each other as: Network Functions can be virtualized and deployed without an SDN being required and vice-versa. But, merging the two technologies will lead to high performance.

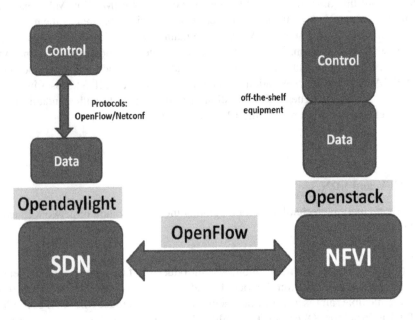

Fig. 3. SDN-NFV relation

Figure 3 shows the intersections between SDN and NFV in terms of proposed virtualization aspect based on some open source projects. In this figure, the SDN can control the NFV Infrastructure through injecting some APIs to the switching devices either hardware or software based like Open Virtual Switch (OVS) [19].

The controller like Opendaylight [20] has two interfaces for communication:

1. The southbound interface which is a standard one using OpenFlow protocol [21] to execute the rules or flows actions inside the OpenFlow devices (OpeFlow-enabled router/switch).
2. The northbound interface is dedicated between the application and the controller. There is no standardization till now for this interface but the majority used APIs for this communication.

Using OpenFlow, The administrator can change any network switch's rules when necessary in a dynamic flow tables or even blocking specific types of packets with a very granular level of control. This is especially helpful in a cloud computing multi-tenant architecture (either VMs or Containers), because it allows the administrator to manage traffic loads in a flexible and more efficient manner. Essentially, this allows the administrator to use less expensive commodity switches and have more control over network traffic flow than ever before. Figure 4 shows some OpenFlow messages captured from Wireshark packet analyzer during our test for an SDN controller (IP: 192.168.0.155) to an OpenFlow client (OVS) (IP: 192.168.0.152) using TCP connection with the standard port: 6633.

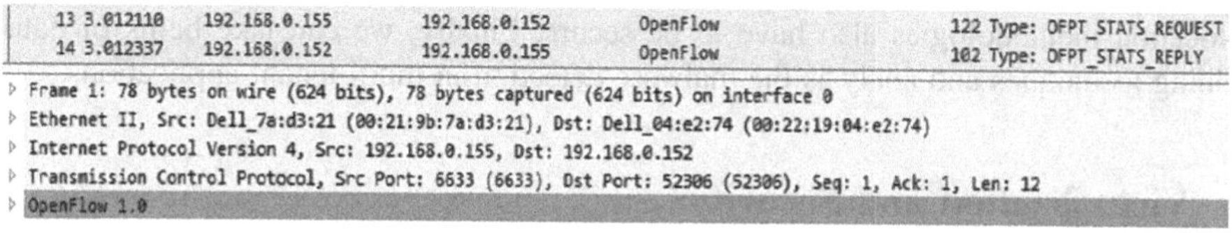

```
 13 3.012110   192.168.0.155      192.168.0.152     OpenFlow          122 Type: OFPT_STATS_REQUEST
 14 3.012337   192.168.0.152      192.168.0.155     OpenFlow          102 Type: OFPT_STATS_REPLY
▷ Frame 1: 78 bytes on wire (624 bits), 78 bytes captured (624 bits) on interface 0
▷ Ethernet II, Src: Dell_7a:d3:21 (00:21:9b:7a:d3:21), Dst: Dell_04:e2:74 (00:22:19:04:e2:74)
▷ Internet Protocol Version 4, Src: 192.168.0.155, Dst: 192.168.0.152
▷ Transmission Control Protocol, Src Port: 6633 (6633), Dst Port: 52306 (52306), Seq: 1, Ack: 1, Len: 12
▷ OpenFlow 1.0
```

Fig. 4. OpenFlow packets encapsulation through wireshark capture.

The OpenStack [22] is one of the famous open source cloud project. It had many nodes which are the outputs of many collaborated projects contributed in OpenStack project.

3.4 SDN Benefits in Virtual Topologies

SDN offers more efficient ways to support virtual topologies over traditional VLANs, because a controller can assign and release resources as needed, whenever and wherever that may be. For instance, rather than configuring specific devices (as required in a VLAN scenario), specific group applications can be set to send the SDN controller a request for bandwidth between group users' affected devices. If the controller determines that any changes are required on the devices, those changes can be pushed out to users as soon as they join a group.

Another thing, with the SDN framework either for VM-based or container-based instances solutions, SDN can isolate those instances from dealing with the low-level networking infrastructure considerations and create a simpler view of the network. For the connectivity, OVS such as layer 2 networking concept is concerned by the network connectivity issues. SDN concerns layer 3 IP addresses for OpenFlow communications. Therefore, it simplifies the network configuration and speeds up the deployment process.

3.5 Virtualization Mapping Issues

The ability to run Virtual machines in different networks and with different security levels is a challenge. Also, virtualization comes with different security issues and problems because it includes many technologies such as networks (different MACs and Ethernets), databases, operating systems, and virtualization realizations either using hypervisors or Docker containers …extra. After the integration of this aspect, the need for privacy and security of data or applications causes that the implementers, developers and research community must re-thinking and re-designing solid and secured solutions and methodologies. Hence, security issues of these systems and technologies are solved as much as possible and became applicable to virtual environments. First of all, the network which used to interconnect the systems in this environment must be secured to mitigate risks and network attacks. Also, mapping of the guests (Virtual Machines) to the hosts (physical machines) have to be performed securely. Moreover, we have to ensure that appropriate policies are enforced for data sharing as security not only involves the encryption of the data. In addition, memory management and resource allocation methodologies also have to be secure. Finally, we can take befits of data mining techniques and analysis the malware detection in multi-tenant applications.

4   Virtualization and Security

4.1   Trusted Platforms

In the context of trusted platforms, trusted platform module (TPM) is a secure chip attached mainboard on the physical platform (such as a PC, a printer, etc.) to keep the integrity of the system. Its specification is created and promoted by the Trusted Computing Group (TCG). According to TCG, TPM is equipped with volatile and non-volatile memory for storing computational variables and secret data. By providing platform monitoring, secure storage, encryption operations and authentication service [23], the TPM is also known as the primitive security for the computing platform. For integrity measurements, TPM uses a special registration that is the PCR (Platform Configuration Register) to store the most important volatile data. The non-volatile memory deals with storing TPM secrets such as security keys. The TPM has two distinguished keys: the Endorsement Key (EK) and the Storage Root Key (SRK). The EK is generated by the manufacturer and unique to the TPM chip. This key is used for TPM identity and is never revealed externally. Related to the EK are Attestation Identity Keys (AIKs). The AIK is a 2048 bit RSA key for signing data that is generated locally by the TPM but may be leaked outside. The latter, Store Root Key, is the first key to be created by the TPM in the initializing process. The SRK is a non migratable key located as the root of TPM's key hierarchy for cryptographic protection of keys held outside the TPM.

With the advantage of virtualization technology as discussed earlier, the convergence between trusted computing and virtualized computing enables a novel security method. In this context, because the physical TPM that is linked to the underlying physical machine cannot attest directly guest OSs, it has been virtualized to provide its functionalities for each VM, which is running on a single platform. This makes VM feel it has its own private TPM. As a result, by virtualizing, a single physical TPM works as the Root of Trust which can be served by multi virtual environments on the trusted platform. For this reason, there are different implementation models for TPM virtualization [24–27]. Table 1 presents the difference between the physical TPM (noted pTPM) and virtual TPM (noted vTPM).

Table 1. pTPM vs vTPM.

Criterion	Physical TPM (pTPM)	Virtual TPM (vTPM)
Design	Hardware (TPM chip)	Software
Resources	Endosement Key (EK), Storage Root Key (SRK), Attestation Identity Key (AIK), and Platform Configuration Registers (PCRs)	vEK, vSRK, vAIK, and vPCRs
Specifications	Standardized by Trusted Computing Group (TCG)	Imitate the functionality of the pTPM
Security	Trust anchor, high security level	Low security level in comparison with pTPM
Operation platform	One to one	Multi to one

4.2 Virtualized Referenced Architecture

On the virtualized reference architecture shown in Fig. 5, the TPM attached to the hardware layer while other security measures attached to the hypervisors sub-layers. In this context, we are going to differentiate between the two realizations either for hardware H/W part in terms of TPM as explained in previous sections or software S/W part in terms of Network Virtualization as will be detailed in the next sections.

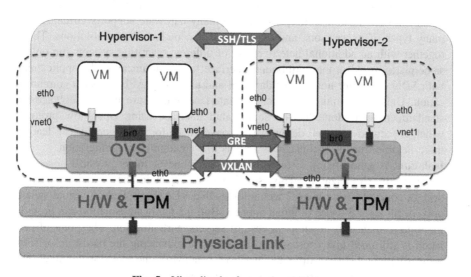

Fig. 5. Virtualized referenced architecture.

Securing VM actions either creation, migration, or cloning is considered as an important step for assuring security in the whole elements (either VMs or processes) and software. As it is a prerequisite in the whole process, we can use one of the following protocols; secure shell (SSH), Transport Layer Security (TLS), or Transmission Control

Protocol (TCP) which are protocols aim to provide privacy and data integrity. OpenSSH is a free version of the SSH connectivity tools, it provides several authentication methods and secure tunneling capabilities for the hypervisors actions. Also, Transport Layer Security (TLS) is the current version of SSL (Secure Socket Layer). TLS uses obviously encryption to encrypt data in transit datagrams.

5 Virtualization Challenges in Summary

Multi-tenant platforms must be highly secured and scalable. So, the main two issues in such kind of platforms are:

- **Security:** Either hardware or software based. Tenants' isolation (either VMs or Containers) is a big challenge either for the developers or the administrators.
- **Scalability:** The virtualized platforms are subject to elasticity principle which means easy provision for the resources in real time. Also, they accommodate huge number of virtual instances run together in same time on same hosts or on different hosts.

There is no doubt that virtualization has been improved the applications delivery and optimized the resources allocation. Meanwhile, it gave the collaborators the shared access for multi-tenant applications that customized on their demands. But, there are many open issues that can be subject to research directions as follows:

- **Heterogeneity:** it is one of the main open issues that required more clarification under the virtualization concept. There are many hardware supports virtualization, many types of hypervisors, and many types of containers based solutions. Those together with an additional heterogeneity of different operating systems run on the same platform. Also, the northbound interface for communicating the applications with SDN controller not yet standardized and used only APIs for such communication. Unified communication for this interface will assure the convergence for standardization as done for the southbound interface using OpenFlow protocol which is secured and well defined. The convergence for heterogeneity can lead to more flexible trusting for instances actions like creation/cloning/migration.
- **Isolation:** isolating many virtual instances over one or many physical hosts is a challenge in virtualization. This isolation can subject to the virtual connecting bridges concepts using layer 2 technologies like VLANs applied on virtual switch ports or till layer 3 based different subnets. But, the hardware isolation can provide more performance evaluations in case of servers' virtualization. The point of isolation is still need many research work either for enhancing the hardware or software isolations between different instances meanwhile the isolation for inter and intra communication a cross virtual data center is a vital issue.
- **Security:** the security of multi-tenant platforms has two concerns; hardware part and software part. Regarding the hardware part, the TPM as explained before with the evolutions of vTPM [24] solutions can improve the trusting for those tenants running. While for software part, there are many assessment tools either for identities, licenses, authentications and VPNs. Moreover, the security of communication

between the hardware TPM and the virtualized ones. This is including the scalability and security issues for such management for tenants especially for the mobility over different platforms.

- **Performance:** the performance analysis either for VMs, Containers or NFVs over virtualized infrastructure is important. The main metrics are the system, hardware and network that affect the overall cost for the deployment and meanwhile for the systems throughput. The work in [28] listed the main issues relevant to network and system for instances migration controlled by classical cloud solutions, NFV and SDN. Moreover, regarding to the performance, many research achievements can be realized in terms of switching speed for OVSs or bridges that communicates the virtual instances over the top of hypervisors. The whole virtualized hardware solutions vSwitch, vCPU, vRAM, vStorage and vNIC for networking are all subject to the performance evaluation.

6   Conclusion

In this paper, we surveyed some technical challenges related to the virtualization. Through the virtualization techniques, there are different issues either concerning hardware or network and software. In general, the virtualization layer assimilated by the hypervisors is the guarantee for everything. While, scalability of instances trusting using either hardware based TPM or virtual TPM is a challenge. Therefore, isolating the created instances in the form of VMs/Containers subjected to the hardware trusting and the software supporting by the virtualization layer. Also, assuring the secure communication for the isolated tenants can be done either using the hardware based solutions assured in the internal paths by the kernels or through tunneling techniques for intra or inter communications. Moreover, the centralized solutions proposed by SDN can improve the policy based access rules applied on the forwarding plane in the network using OpenFlow protocol.

In our future work, we aim to identify the correlation between hardware and software to improve the performance of certifying the trusting of virtual instances, to test the overall performance in terms of distributed solution based SDN for securing the created/migrated instances over different platforms, and finally to construct dynamic policies that can assure the flexibility for heterogeneity either in software or hardware based tenants' isolation.

References

1. Ahmad, R.W., Gani, A., Hamid, S.H.A., Shiraz, M., Yousafzai, A., Xia, F.: A survey on virtual machine migration and server consolidation frameworks for clouddata centers. J. Netw. Comput. Appl. **52**, 11–25 (2015)
2. Aguiar, A.D.C.P.D.: On the virtualization of multiprocessed embedded systems. Ph.D. dissertation, Pontifícia Universidade Catolica do Rio Grande do Sul (2014)

3. Mijat, R., Nightingale, A.: Virtualization is coming to a platform near you. ARM white paper (2011)
4. Le Vinh, T., Bouzefrane, S.: Trusted platforms to secure mobile cloud computing. In: The 16th IEEE International Conference, pp. 1096–1103, August 2014
5. Liang, C., Yu, F.R.: Wireless network virtualization: a survey, some research issues and challenges. IEEE Commun. Surv. Tutor. **17**(1), 358–380 (2015)
6. Jain, R., Paul, S.: Network virtualization and software defined networking for cloud computing: a survey. IEEE Commun. Mag. **51**, 24–31 (2013)
7. Popek, G.J., Goldberg, R.P.: Formal requirements for virtualizable third generation architectures. ACM SIGOPS Oper. Syst. Rev. **17**, 412–421 (1974)
8. VMWare Inc.: Introducing VMware virtual platform. Technical white paper, February 1999
9. Understanding Full Virtualization, Para-virtualization, and Hardware Assist. VMWre white paper (2007)
10. Rodriguez-Haro, F., Freitag, F., Navarro, L., et al.: A summary of virtualization techniques. In: The 2012 Iberoamerican Conference on Electronics Engineering and Computer Science, pp. 267–272 (2012)
11. http://lwn.net/Articles/531114/
12. http://en.wikipedia.org/wiki/Cgroups
13. https://tools.ietf.org/html/rfc5556
14. NFV White Paper: Network Functions Virtualisation, An Introduction, Benefits, Enablers, Challenges & Call for Action, issue 1, October 2013
15. ETSI GS NFV 002 (V1.2.1): Network Functions Virtualisation (NFV); Architectural Framework, December 2014
16. ETSI GS NFV 003 (V1.2.1): Network Functions Virtualisation (NFV); Terminology for Main Concepts in NFV, December 2014
17. Mijumbi, R., Serrat, J., Gorricho, J., Bouten, N., De Turck, F., Boutaba, R.: Network function virtualization: state-of-the-art and research challenges. IEEE Commun. Surv. Tutor. **PP**(99), 1 (2015)
18. Hu, F., Hao, Q., Bao, K.: A survey on Software-Defined Network (SDN) and OpenFlow: from concept to implementation. IEEE Commun. Surv. Tutor. **16**, 2181–2206 (2014)
19. http://openvswitch.org/
20. https://www.opendaylight.org/
21. McKeown, N., Anderson, T., Balakrishnan, H., et al.: OpenFlow: enabling innovation in campus network. ACM SIGCOMM Comput. Commun. Rev. **38**, 69–74 (2008)
22. https://www.openstack.org/
23. TPM Main Specification. http://www.trustedcomputinggroup.org/resources/tpm_main_specification
24. Perez, R., Sailer, R., van Doorn, L., et al.: vTPM: virtualizing the trusted platform module. In: Proceedings of 15th Conference on USENIX Security Symposium, pp. 305–320 (2006)
25. England, P., Loeser, J.: Para-virtualized TPM sharing. In: Lipp, P., Sadeghi, A.-R., Koch, K.-M. (eds.) Trust 2008. LNCS, vol. 4968, pp. 119–132. Springer, Heidelberg (2008). doi:10.1007/978-3-540-68979-9_9
26. Krautheim, F.J., Phatak, D.S., Sherman, A.T.: Private virtual infrastructure: a model for trustworthy utility cloud computing. In: DTIC Document (2010)
27. Strasser, M., Stamer, H.: A software-based trusted platform module emulator. In: Lipp, P., Sadeghi, A.-R., Koch, K.-M. (eds.) Trust 2008. LNCS, vol. 4968, pp. 33–47. Springer, Heidelberg (2008). doi:10.1007/978-3-540-68979-9_3
28. Ibn-Khedher, H., Abd-Elrahman, E., Afifi, H.: Network issues in virtual machine migration. In: ISNCC, IEEE, Hamamet, Tunisia, vol. 1, pp. 1–6, May 2015

Prediction of a Mobile's Location Based on Classification According to His Profile and His Communication Bill

Linda Chamek[1,2], Mehammed Daoui[1], and Selma Boumerdassi[2(✉)]

[1] LARI, University Mouloud Mammeri, Tizi-Ouzou, Algeria
{chameklinda, mdaoui}@yahoo.fr
[2] Conservatoire National des Arts et Métiers CNAM, Paris, France
Selma.boumerdassi@inria.fr

Abstract. In this paper, we present a new approach to predict the displacement of a mobile based on classification according to profile (all significant information that characterizes a user), and taking account of communication bill of this one. Our solution can be implemented in a third generation network, by exploiting information of users (age, function, residence place, work place ...), the existing infrastructure (roads ...) and the historical of displacements.

Keywords: Mobile network · Prediction · Profile · Data mining

1 Introduction

New technologies of mobile networks open to users a high perspective. In addition to traditional communication, they provide a wider bandwidth and they allow real time transmission of data. They also allow integrating multi-media applications such as: interactive games, telephony IP ... etc.

These new applications which require high quality of service (QoS) must have the guarantee of a good execution without delay, discontinuity or brusque interruptions which happen during the mobiles displacements, when they move from a cell to another. We call it handoff.

The mobility prediction can improve QoS by intervening in several functions of mobility management, such as mobiles' localization.

To convey calls and data to a mobile, the system sends paging messages in the network to locate it.

These messages consume bandwidth. If network can know in advance the itinerary that the mobile will follow during its displacement, it will be able to search it in a limited number of cells, reducing, therefore, the number of paging messages and the search time.

In this paper, we present a prediction strategy of displacements based on the classification of users according to their profiles. First we identify two classes of users

S. Boumerdassi et al. (Eds.): MSPN 2016, LNCS 10026, pp. 63–75, 2016.
DOI: 10.1007/978-3-319-50463-6_6

according to their communications bill. For a privileged user, the model classes him according to his profile. It examines old displacements of known users, having the same profile as him, to predict his future location.

The rest of the paper is organized as follows. In Sect. 2, we present a state of art of different techniques of prediction, in Sect. 3 we present the importance of the use of data-mining in mobility prediction, in Sect. 4 we propose a prediction solution and in the last section, we present the evaluation of the solution.

2   State of Art

Several techniques allowing prediction have already been discussed.

One of the most used techniques rests on the localization by GPS (Global Positioning system). The mobile sends its position obtained by GPS to its base station. The latter determines if the mobile is at the edge of its cell. At each reception of the position of the mobile, the system calculates the distance separating the mobile from the neighboring cells, and the shortest (near) distance is selected [1].

Hsu et al. [2] suggests a solution based on the definition of a reservation threshold. The idea is to compare the signal received by the mobile coming from the neighboring cells. If this signal is lower than the threshold, it is concluded that the mobile moves towards this cell. In [3], a map of signal power is maintained by the system. It represents the various signals recorded in various points of the cell. They use this map to know the position of the mobile, and to extrapolate his future position.

In [4, 5], mobility rules are generated based on a history of the movements that each mobile built and maintains during its displacement. These rules are used in the prediction process. In fact, it was observed that the users tend to have a routine behavior. Knowing that, and knowing the usual behavior of the users, it becomes possible to deduce the next cell which a user will visit.

These solutions are limited because they are based on either a probabilistic model which does not completely reflect the users' behavior, or on the user individual history which can be missing or insufficient.

The authors propose a technique based on the use of a multi-layer neural network in [6] to exploit the history of the mobile movements. The recent movements of the mobile are initially collected in order to know in which LA (location area) it is. A mobility model of the users is initially processed, and then it will be injected into the neural network. The mobility model represents the mobile movements history recorded in an interval of time. The movement is defined in terms of the taken direction, and the distance covered. The role of the neural network is to capture the unknown relation between the last values and future values of the mobility model; that is necessary for the prediction.

The disadvantage of this method is that they require a long training phase on mobile user behavior before the prediction succeeds. Moreover, the mobile user can change his behavior during the training phase or can go to a location he has never visited before, thus making the prediction ineffective.

The authors in [7] propose an algorithm which spread out in three phases. The first consists in extracting the movements of the mobile to discover the regularities of the inter-cellular movements; it is the mobility model of the mobile. Motilities rules are extracted from the preceding model in the 2nd phase. And finally, in the 3rd phase, the prediction of mobility is accomplished by using these rules.

A method called DCP (Dynamic Clustering based Prediction) is presented in [8]. It is used to discover the mobility model of the users from a collection containing their trajectories. These rules are then used for the prediction. The trajectories of the users are grouped according to their similarities.

In [9], Samaan and Karmouch present a solution which includes spatial and user contexts. The spatial context consists of a set of abstract maps describing the geographical environment in which the mobile user progresses. Places, buildings and roads which lead to these places are described in these maps. The user context includes a set of information related to the mobile user making it possible to predict his mobility. This information is then combined by using Dempster Shafer algorithm to predict the future mobile location. Even if this solution seems to provide suitable results when mobile history is lacking, it is however too constraining because it requires additional information not easy to acquire and likely to frequent change.

Daoui *et al.* [10, 11] presents a technique of prediction based on the modeling of mobile displacements by an ants system. This model allows the prediction based on old displacements of the mobile and those of the other users who go in the same direction.

The author in [12] presents a technique of prediction based on classification of users according to their profile (all their personal information like age, sexes, place of work, ...) a new user is compared to all other users in cell and put in a class, then the history of displacements of the users in this class is used for predicting the next cell of the new user.

Chen et al. propose in [13] a solution using channel state information CSI and handover history. The prediction is made by solving a classification problem via supervised learning. The overall scheme consists of two phases, training phase and prediction phase.

In the first phase, they inject in the system a set of training sequences CSI, their associated previous cells and their next cell indices. A learning algorithm derives from these data several classifiers corresponding to all the possible previous cells. The prediction phase selects a specific classifier to match a previously unknown sequence CSI to a predicted index of the next cell.

This solution seems to provide suitable results; however it requires a long learning phase and handling a lot of information: the speed, trajectory, the channel information, paths to take. It takes time for the treatment of all these data.

In [14], the authors present a new approach that considers the temporal aspect in the prediction. They start by a long learning phase in which they collect information about subjects and their call detail, they sort them and they select the most relevant.

In the prediction phase, localization traces is collected and divided on clusters, they also built clusters of times such that each cluster contains time intervals which show similarities in the probability distribution of locations. The prediction is made by basing on the estimate of the maximum likelihood.

In this solution authors take a long time for collecting and processing all information they need, and they do not take the individual aspect of users when they built clusters of locations.

Zhang et al. propose a technique that aims to predict the position of a user during a time interval of one to six Hours [15]. For this, three components are used: periodicity predictor, social interplay predictor and self adjust learning.

The first component is used to manipulate the start of regularities of user mobility from spatial and temporal perspective; it exploits the periodic behavior of the users.

The inputs of the system are the historical traces of an individual GSM user. For traces GSM of users used as inputs, social relations are first discovered between two arbitrary users. A relevant time list containing the possible locations traversed by the user is constructed. The list is sorted by the probability that the user moves to the corresponding cell in order to meet a friend (User).

Two lists are generated from the two previous phases. The last component aggregates these lists and deduces the future cell.

3   Data-Mining and Prediction

The mobile's displacements are often generated by socio-economic needs and are governed by the topography of the roads and infrastructures covered by the various cells of the network such as: schools, factories, supermarket, highway, etc. The displacements related to the socio-economic needs are usual, and consequently, represent a regular aspect.

A study [16] carried out in the USA for better organizing public transport has shown that nearly 80% of users displacements relate to work and nearly 20% relate to social or cultural reasons. During holidays, the percentage is only nearly 2%. It has also shown that displacements are influenced by the infrastructures of the places (trades, highways, streets, paths... etc.). Displacements for work and social reasons are the most frequent and the most usual. The knowledge of the history (habits) of a user and his current location (on a road for example) could be useful to determine his probable future location.

Information concerning a user and characterizing it, in other words its profile, are also of great importance. In fact, knowing certain characteristics of a user helps us to know his future displacements with a great probability. For example, a person of an age ranging between 18 and 25 years, who is student, will be located probably in the campus one day of week. The people having important incomes make their purchases in luxury shops, contrary to the others which prefer supermarkets. The profiles of mobility of these people are thus different (Fig. 1).

The knowledge of the profile of network's users can help us to predict their displacements. Group the mobiles according to their profiles allows to exploit mobility information of other mobiles to predict the future position of a user.

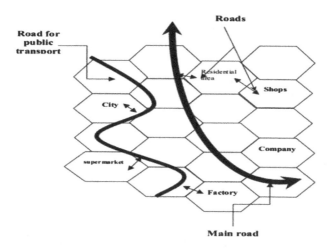

Fig. 1. Example of displacement according to user's profile

Many definitions of the data-mining can be found so this domain is the subject of research. Engineers, statisticians, economists, etc., can have different ideas on what this term recovers. We retain a definition which seems to make the compromise between various designs. We can define data-mining by the process allowing the extraction of predictive latent information from wide database [17].

Classification consists of assigning an object to a certain class based on its similarity to previous examples of other objects.

Classification reach to predict the class of a new user based on the class of users who are in data base. In our case, it is to predict the future cell of a user based on his last displacements and these of other users who have the same profile. The basis of this idea is that displacements of users are often regular and individuals of the same profile perform similar movements.

4 Prediction Solution

We suppose an architecture of third generation network composed of a set of cells. Every cell is generated by a base station. The base stations are connected to the core network wired backbone (Fig. 2). We assume that the core network has personal and professional information about users such as age, marital status, occupation etc. This information may be collected when subscribing to network services. We also assume that each base station has a history of movements of mobile users.

We propose a solution of prediction based on the use of a technique of datamining which is classification.

This approach allows us to consider the prediction from a user perspective, not only the network side. Indeed, in our approach, we consider the specific character of each user (its profile) and their mobile behavior (their historical mobility). Unlike other

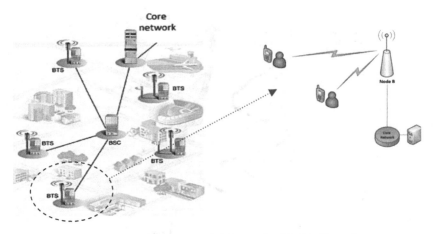

Fig. 2. Architecture of a Third Generation Mobile Network

techniques presented in the Literatures most of which do not take into account the individuality of the users.

This point is important because the mobile behavior of users (move) depends on personal characteristics. (Example: a 20 year old student will not move in the same way that an adult 45/50 years) because these people do not have the same centers of interest and therefore they do not necessarily go to the same places.

Basing on the profile of the users, we can deduce their habits and thus their displacements.

If a new user enters the network and we don't have his movement history (we don't know his habit of displacements), we can always use the history of the other users present in the network having the same profile as him (the person having the same profile adopt the same behavior).

The Classification Technique, of datamining, used in our approach allows identifying classes of individuals that are alike (the notion of profile). With this principle of similarity, persons belonging to the same class are more likely to behave in the same way, that is to say, have the same interests, and have the same mobile behavior in general.

We propose initially to pre-classify users into two classes according to their communication bill. In fact, this will facilitate us the classification according to profile and prediction. We form two great classes: large and medium consumers. Once this distinction made, when classifying a user according to the profile, we target individuals with which we will classify him.

A person communicating much will look like more to individuals belonging to the 'large consumers' class than to those of 'low consumptions'.

4.1 User's Profile

The profile of a user is all useful information allowing to the system to understand his behavior and all information's characterizing him such as: age, sex, profession, work place, place of residence, … etc. These information can be recovered when the user subscribe.

4.2 Memorizing Displacements

History of mobility of a user consists to save the various transitions carried out between the various cells from the network during its displacement. These information can be recovered in the log files of the base stations. Each base station maintains a history of various displacements of the mobiles having visited the cell. The structure of a line of history is presented in Table 1.

Table 1. Structure of a history line

ID mobile	Source cell	Destination cell	Date

– Mobile Id: unique mobile identifier
– Cell source: indicates the cell from which the mobile came
– Cell dest: indicates the cell towards which the mobile is going
– Moment: indicates the moment of displacement (workday or bank holidays)

The last parameter is important. The displacement of user can be different according to this parameter. For example a user goes in a relaxation place on the weekend instead of going to work.

4.3 Prediction of Localization

For predict displacement of the mobiles, we propose a solution which proceeds in two principal stages:

In the first stage we start with a pre-classification. We identify two classes of users according to their communications bill:

• Large consumers' class which contains all users who use their mobiles frequently (it contains persons like students, businessman, traders, …., etc.)
• Medium consumers class which has low priority (it contains senior person for example)

The second part consist to classify a new mobile, that is pre-classified, according to its profile, and exploit the historical of movements of its neighbors having the same profile as him to predict its future cell [12].

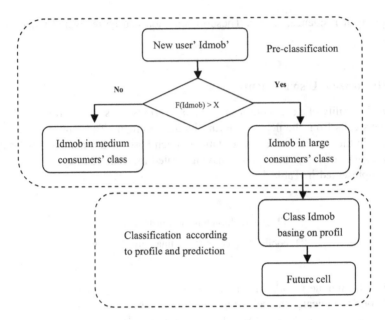

Fig. 3. Diagram of the proposed solution

Procedure Pre-classification

```
Begin
Int x (value defines a threshold)
F= bill (Idmob)
If F> x then
    Put Idmob in the large consumers's class
    Predict his future cell
Else
Idmob has low priority
```

Classification is used for predict the future cell of a mobile user x. We select N mobile users who are already located in the same cell. We compare the user x to every individual y of N users using distance D [18].

$$D(X,Y) = \sqrt{\sum_{i=1}^{m}(x_i - y_i)^2} \qquad (1)$$

With:

- X_i and y_i are values of attributes of individuals X, Y (like age, function, ...)
- m is the number of known attributes of individual

So, we select K individuals nearest to X. Then we take displacement history of these K individuals. The destination cell which is the most frequented is considered the future predicted cell.

The algorithm proposed is summarized as follows:

Let I= {Y1, Y2,... Yn} be the set of the N individuals being in cell C

Entry: Let X be a new individual for whom we want to predict the future cell.

Parameter K corresponds to the number of nearest neighbors to take into account

 Parameter L corresponds to the number of history lines to take for each nearest neighbor

Exit: the future cell to predict

Algorithm:

For (j = 1 to N) **Do**

1. Calculate the distance between Yj and X of the cell C d(X, Yj)

2. Record this distance in the vector tab

Done

3. Sort the calculated distances (the vector)

4. Select K smaller distances,

5. Select the history of K individuals closest to X

6. Determine the most frequent destination cell and return it as the future cell to predict

5 Adjustment and Evaluation

Most data mining algorithms require training phase for adjusting their parameters. In the case of classification it provides the best value of two parameter K (number of nearest neighbor to take into account) and L (number of line of history to take by close neighbor).

In the study of the mobility management and in the absence of a real trace of mobiles displacement, we can resort to a model. The choice of a realistic mobility model is essential.

This model reproduces, in a realistic way displacements of a set of users within the network. The majority of works presented in the literature use probabilistic models (Markov model, poisson process, etc.) which generate either highly random displacements or highly deterministic displacements which do not reflect the real behavior of the mobile users.

In our approach, we have chosen the activity model presented in [19]. This model is based on the work carried out by planning organizations and uses statistics drawn from five years observation on users' displacements. It simulates a set of user's displacements during a number of days. Generated displacements are based on each user's activity (work, study, etc.), the locations of these activities (house, work places, schools) as well as the ways which lead to these locations.

The simulator rests on the statistics of displacement led in the area of Waterloo [20] and recorded in the form of matrix called activity matrix indicating the probability of arrival of an activity, and duration matrix indicating the probability that an activity takes a given period. These statistics as well as information concerning the users like the profile (full time employee, student, part-time employee, etc.) and the infrastructures (roads, trade, stadium, etc.) are recorded in the simulator database.

The area of Waterloo is divided into 45 cells as indicated in Fig. 3. According to the activity of the user, the simulator generates an activity event for a user based on the activity matrix and assigns it to a cell. It generates then the displacements relative to this activity before generating the following activity. The process continues until the end of the simulation.

5.1   Adjustment and Evaluation of the Classification Algorithm

The adjustment of the classification algorithm consists of determining optimal values for the two parameters K and L. K is the optimal number of neighbor to consider, and L is the number of history lines to take for each near individual. The evaluation is done based on rate prediction which is the ratio of the number of correct predictions to the total number of attempts to predict.

Figure 4 shows the ratio prediction in function of the parameter K, with the L value fixed to 4 (take 4 lines of history for each neighbor). The prediction ratio rises to stabilize at K = 30 with ratio prediction of 60%.

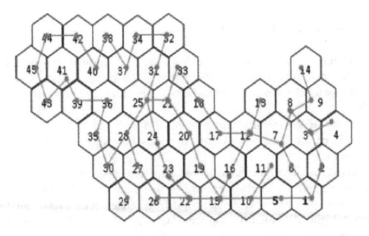

Fig. 4. Cellular structure of the simulator

Figure 4 gives the ratio prediction as a function of the parameter L when K is fixed to 30 (the value that we get above). The prediction ratio rises with the rise of the value of L. A better ratio is obtained for L = 45 with a ratio of 70%. So, we can only keep the last 45 displacements of neighbors' users (Figs. 5 and 6).

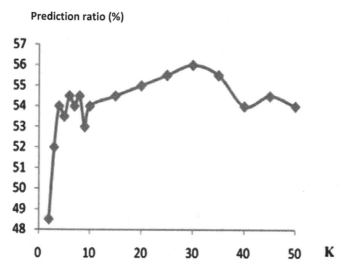

Fig. 5. Prediction ratio according to parameter K (optimal number of neighbor)

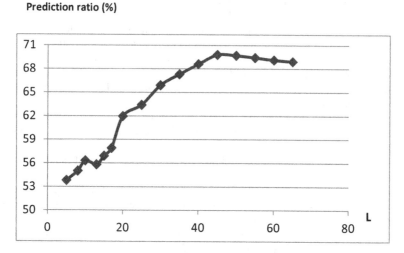

Fig. 6. Prediction ration according to parameter L (number of history lines to take for each near individual)

6 Conclusion

Human displacements are often caused by socio-professional needs. They are linked to existing infrastructure (roads, transport, workplace location … etc.). It is therefore possible to predict the future position by looking for links between these movements and other available information such as user profiles and the location of the infrastructure.

Due to the complexity of the characteristics of human mobility and the absence of reliable mobility rules for prediction movements, data mining can be a solution to the problem of prediction. Both techniques presented in this paper show that it is possible to predict 70% of the movements of mobile users.

References

1. Zhuang, W., Chua, K.C., Jiang, S.M.: Measurement-based dynamic bandwidth reservation scheme for handoff in mobile multimedia networks (1998)
2. Hsu, L., Purnadi, R., Wang, S.S.P.: Maintaining quality of service (QoS) during handoff in cellular system with movement prediction schemes. In: IEEE (1999)
3. Choi, S., Shin, K.G.: Predictive and adaptive bandwidth reservation for hand-offs in QoS-sensitive cellular networks. In: IEEE (1998)
4. Shen, X., Mark, J.W., Ye, J.: User mobility profile prediction: an adaptive fuzzy interference approach. Wirel. Netw. 6, 362–374 (2000)
5. Ashrook, D., Staruer, T.: Learning significant locations and predicting user movement with GPS. In: Proceedings of the 6th International Symposium on Wearable Computers TSWC 2002 (2002)
6. Soh, W.S., Kim, H.S.: QoS provisioning in cellular networks based on mobility prediction techniques. IEEE Commun. Mag. 41, 86–92 (2003)
7. Liou, S.C., Lu, H.C.: Applied neural network for location prediction and resource reservation scheme in wireless network. In: Proceedings of ICCT (2003)
8. François, J.M., Leduc, G.: Entropy-based knowledge spreading and application to mobility prediction. In: ACM CoNEXT 2005, Toulouse, France, 24–27 October 2005
9. Samaan, N., Karmouch, A.: A mobility prediction architecture based on contextual knowledge and conceptual maps. IEEE Trans. Mob. Comput. 4(6), 537–551 (2005)
10. Daoui, M., M'zoughi, A., Lalam, M., Belkadi, M., Aoudjit, R.: Mobility prediction based on an ant system. Comput. Commun. 31, 3090–3097 (2008)
11. Daoui, M., M'zoughi, A., Lalam, M., Aoudjit, R., Belkadi, M.: Forecasting models, methods and applications, mobility prediction in cellular network, pp. 221–232. i-Concepts Press (2010)
12. Chamek, L., Daoui, M., Lalam, M.: Mobility prediction based on classification according to the profile. Journées sur les rencontres en Informatique R2I, 12–14 june 2011
13. Chen, X., Meriaux, F., Valentin, S.: Predicting a user's next cell with supervised learning based on channel states. In: IEEE 14th Workshop on Signal Processing Advances in Wireless Communication (SPAWC) (2013)
14. Gatmir-Motahari, S., Zang, H., Reuther, P.: Time-clustering-based place prediction for wireless subscribers. IEEE/ACM Trans. Netw. 21(5), 1436–1446 (2013)

15. Zhang, D., Zhang, D., Xiong, H., Yang, L.T., Gauthier, V.: NextCell: predicting location using social interplay from cell phone traces. IEEE Trans. Comput. **64**(2), 452–463 (2015)
16. Hu, P., Young, J.: 1990 Nationwide Personal Transportation Survey (NPTS), Office of Highway Information Management, October 1994
17. Han, J., Kamber, M.: Data Mining Concepts and Techniques, 2nd edn. Morgan Kaufmann, San Francisco (2006)
18. Wu, X., Kumar, V., et al.: Top 10 algorithms in data mining. Knowl. Inf. Syst. **14**(1), 1–37 (2008)
19. Scourias, J., Kunz, T.: An activity-based mobility model and location management simulation framework. In: Proceedings of the Second ACM International Workshop on Modeling, Analysis and Simulation of Wireless and Mobile Systems (MSWiM), Seattle, Washington, USA, pp. 61–68 (1999)
20. Scorias, J., Kunz, T.: A dynamic individualized location management algorithm. In: 8th IEEE International Symposium on Personal, Indoor and Mobile Radio Communications. Waves of the Year 2000, PIMRC 1997, Helsinki, pp. 1004–1008, September 1997

A Comparative Study of the Mobile Learning Approaches

Sameh Baccari[1]([⊠]), Florence Mendes[2], Christophe Nicolle[2],
Fayrouz Soualah-Alila[2], and Mahmoud Neji[1]

[1] MIRACL, University of Sfax, Sfax, Tunisia
sameh_baccari@yahoo.fr, Mahmoud.neji@fsegs.rnu.tn
[2] LE2I, UMR CNRS 6306, University of Bourgogne, Dijon, France
{florence.mendes,fayrouz.soualah-alila}@checksem.fr,
cnicolle@u-bourgogne.fr

Abstract. With the emergence of mobile devices (Smart Phone, PDA, UMPC, game consoles, etc.), and the growth of offers and needs of a company under formation in motion, multiply the work to identify relevant new learning platforms to improve and facilitate the process of distance learning. The next stage of distance learning is naturally the port of e-learning to new mobile systems. This is called m-learning (mobile learning). Because of the mobility feature, m-learning courses have to be adapted dynamically to the learner's context. Several researches addressed this issue and implemented a mobile learning environment. In this paper, we compare a list of mobile learning architectures with methods presented in the literature. The evaluation presents a set of criteria specifically identified to qualify m-learning architectures dedicated to the context-change management.

Keywords: Mobile technology · E-learning · M-learning · Context-change management · Learning method

1 Introduction

The introduction of mobility in a learning process induces new practices and uses which change the conception paradigm of a learning courses. This paradigm was implicitly based on the unity of place. With a mobile learning systems, the learner can continue his education out of a classroom, move during the learning process and changed place using a mobile device whether a Smartphone or a tablet. This mobility led to the appearance of a new paradigm where the learning process has to be dynamically adapted according to the change of the learner's context.

M-learning objects can take many forms, such as text, audio or video documents organized into comprehensive training programs adapted to mobile devices. While there are a large number of courses available on mobile devices, this type of training is still at an early stage. In fact, it is sometimes difficult to adapt of the mobile devices to the available contents in e-Learning. For instance, the problems to migrate e-learning training course to mobile learning systems are not only limited to technical issue, such as the limitation of screen size and the bandwidth, but also to the management of

S. Boumerdassi et al. (Eds.): MSPN 2016, LNCS 10026, pp. 76–85, 2016.
DOI: 10.1007/978-3-319-50463-6_7

change within the context and its impact on the training course. In addition, even if some contents, such as audio and video media course materials, are ideally suitable for mobile use, the existing systems are still unsatisfactory for the users' needs. Our goal is to identify how the existing mobile-learning methods resolve the technical hetero-geneity and bridge the gap change management during the mobile learning process.

The remainder of this article is organized as follows: In Sect. 2, we give a brief definition of the mobile learning. The context and the opportunities of context-aware applications are described in more detail in Sect. 3. Then, in Sect. 4, some existing works in the literature are presented. Then, a comparison between the different architectures proposed in these works is made using a list of significant criteria. Finally, we end up our paper with a conclusion.

2 M-learning

M-learning is described by the use of mobile and wireless technologies allowing anyone to access information and learning materials at any time regardless of the place. Some approaches consider that mobile learning is simply an extension of the E-learning. However, they do not take into account the mobile device limitations, the particular circumstances of mobile learning and the added value of mobility, such as informal learning, learning on demand, in context, through contexts, etc. Mobile learning has been defined as the process of learning and teaching that occurs with the use of mobile devices providing flexible on-demand access (without time and device constraints) to educational resources, experts, peers and services from any place [1].

This evolution of learning can be characterized by the following changes: distance in e-learning and the consideration of mobility with m-learning and omnipresence with the ubiquity ubiquitous learning (ubiquitous learning, pervasive learning). These changes reflect the impact of computer technology, such as mobile and ubiquitous computing, on the learning process.

The shift from e-learning to mobile learning has given rise to much debate among researchers. For instance, Sharma noticed that the shift from e-learning to mobile learning is accompanied by a change in the terminology [2], as shown in Table 1.

Table 1. Comparison of e-learning terminology and m-learning (according to [3])

E-learning	M-learning
Computer	Mobile
Bandwidth	GPRS, 3G, Bluetooth
Multimedia	Objects
Interactive	Spontaneous
Hyperlinked	Connected
Distance learning	Situated learning
More formal	Informal
Simulated situation	Realistic situation
Hyper learning	Constructivism, situationism, collaborative

The main benefits of mobile learning for education and learning are reported as follows [4]: (a) it enables on-demand access to learning resources and services as well as instant delivery of notifications and reminders, (b) it offers new learning opportunities that extend beyond the traditional teacher-led activities and classroom-based ones, (c) it encourages learners to participate more actively in the learning process by engaging them to authentic and situated learning embedded in real-life context and (d) supports on-demand access, communication and exchange of knowledge with experts, peers and communities of practice.

The major difference between learning somewhere on a stationary desktop computer and learning with mobile devices is the context. In fact, mobile devices feature some functionality to capture some background information that can be helpful to personalize the learning experience. When considering mobility from the learner's point of view rather than the technology's, it is more important to say that m-learning is about people moving through environments, their learning as they go, using electronic devices that enables connectivity to information sources and communicating while they are able to change their physical location. In short, our new definition of mobile learning is "context-aware in mobile learning" which discussed in more detail in the next sections.

3   Context-Aware in Mobile Learning

In a mobile learning experience, each learner has to be treated in a different way according to the current situation in which he is learning, e.g. his pre-knowledge or the specifications of the device he is using. Those different conditions are called the context in which the learner is situated.

As mobility is related to mobile learning, mobile devices, capacity, connectivity, user and the environment can all change over time and place. That is to say the set of learning exchange or the learning context can change all the time. A mobile learning's challenge is to exploit applications that can dynamically adapt to different learning situations. M-learning makes learning across contexts: "mobile learning is not just about learning using mobile devices, intended learning across contexts" [5]. Here the focus is on how learners are formed through places and transitions between different contexts.

It should not be a break in learning between the face-and outside. Learning is based on the business continuity through space and time interacting with mobile and fixed technologies. Learning should not be limited to certain environments, but should increase the mobility of the learner through these. With the emergence and evolution of new mobile technologies, adaptation to context has become an indispensable nature of new computer systems for mobile use. It is therefore necessary in the case of an adaptation to the context in learning, determined by the context of the learner what content to send, how, on what tool, etc. The whole learning process has to adapt to these changes in context. On the other hand, learning through contexts requires a series of organized activities, that is to say that learning takes place in a particular context depends on those who were before. This requires that the system takes into account the history of learning to provide the learner with meaningful learning activities and thus to monitor its activities through contexts.

4 Previous Works

In this section, we will review the work existing in the literature by introducing representative approaches and mobile learning platforms. We are interested in the adaptation management which is a very important parameter in managing and customizing the learning resources to the learners. We will focus, also, on the context, which is a central core of all the mobile learning systems.

MOBILearn is a European research and development project which aims at exploring the use of mobile environments to foster informal learning, learning through problem solving and learning at work. As part of this project, a new architecture for mobile learning was constructed. It can help generate contents and services to accompany a learner during his learning activities in a gallery or a museum [6]. In this context, Learning is backed by a set of activities at the museums. Being placed in front of a painting, the visitors can, then, use a PDA or Smartphone to get the relevant information while observing the painting in order to learn. A learner visiting a museum, for the second time, can acquire information related to his previous visit.

E-Bag (virtual bag) is part of the iSchool project for nomadic and mobile learning [7] for which mobility and context are two key elements. Briefly, the iSchool project vision is used to develop a software infrastructure, graphical interfaces and spatial concepts in an interactive environment. The idea behind eBag is the creation of a "virtual school bag" for each student to help him learn through contexts by moving to specific locations (classrooms, laboratories, workshops, libraries, museums, cities, clubs and home). Therefore, the objective of the system is to serve as a "personal and digital warehouse" in which all the resources (texts, photos, videos, etc.) can be stored for internal and external use of the school environment.

The MoULe project (Mobile and Ubiquitous Learning) aims at allowing pupils to use mobile devices in order to build up collaborative knowledge and incorporate learning activities in the classroom and laboratory for situation learning. The MoULe is an environment that helps the users edit and share documents and concept maps using desktop computers and smart phones equipped with GPS. These tools enable students to collect textual content, images, videos and audio recordings while visiting an outside site during their learning activities. Besides, the system allows them to comment on the media they collect and classify it, so that their research and the re-use of the information in collaborative activities will be easier [8].

The mCALS project (mobile Context-aware and Adaptive Learning Schedule) is a context-aware mobile learning system developed for supporting Java programming learning. The goal of the system is to select appropriate learning objects for learners based on their current context and preferences. User context attributes include their location, and user preference attributes include their knowledge level for the topic (in this case Java) and their available time. The system is made up of three layers: Learner Model Layer, Adaptation Layer and Learning Objects (LO) Layer. Learner model layer collects, organizes and manages the learner's context, which can characterize the learning situation for learning objects adaptation. Adaptation layer is in charge of selecting appropriate learning objects based on the learner's current context with a series of adaptation mechanisms. Learning objects layer stores and manages learning objects in a learning object repository, with which learners are provided during their learning [9].

Nguyen Pham and Ho present CAMLES (Context-Aware Mobile Learning English System) to help students learn English as a foreign language to prepare them for TOEFL test by suggesting topics they need to learn on the basis of their test results. The test provides an adaptive content for different learners in context including location, time and the learner's knowledge. The system architecture of CAMLES includes three layers: Context Detection Layer, Database Layer and Adaptive Layer. The Context Detection Layer identifies the context factors such as location, time interval, manner of learning and learner's knowledge that impact selection of adapted learning contents for different learners. The Database Layer consists of context data, content data, the learner's profile and test. The Adaptive Layer includes an adaptive engine which selects learning contents according to the learner's learning context based on a set of if-then rules stored in the rules repository [10].

Yin proposed the design for a contextual mobile learning system known as SAMCCO (French abbreviation for "contextual and collaborative mobile learning system for professional fields") [11]. It is based on EPSS (Electronic Performance Support System) whose goal is to group storage of technical, working and learning data in order to provide not only just-in-time and just enough training, but also information as well as tools and help mastering or repairing equipment, appliances or products disseminated in the smart city environment. This system is able to bring relevant information designed to maintain or ensure appropriate performance of smart city users whenever and wherever needed, thereby enhancing the performance of the industry and the company as a whole. EPSS is used to store and deliver plant reference materials including: training documents, operating procedures and historical maintenance information. SAMCCO edits and organizes learning contents stored in the EPSS information database, which is an essential professional learning resource offering abundant and well-structured learning contents.

In 2013, SoualahAlila et al. [12] proposed an approach for context-based adaptation for m-earning known as CAMLearn (Context-Aware Mobile Learning), making use of learning practices already deployed in e-learning systems and adopting them in m-learning. This system is built around an ontology that both defines the learning domain and supports context-awareness. The use of this ontology facilitates context acquisition and enables a standard-based learning object metadata annotation. It, also, uses a set of ontological rules to achieve personalized context-aware learning objects by exploiting knowledge embedded in the ontology. The future adaptive system will offer an optimized panel of learning objects matching with the learner's current context. CAMLearn consists of two parts: the first part consists of a knowledge server where data and business processes are modeled by evolutionary ontology and business rules, and the second part is based on metaheuristics algorithms allowing analyzing business rules and ontology to allow a good combination of learning content.

The UoLmP project (Units of Learning mobile Player) Project [13] is intended to present an adaptive, personalized and context sensitive mobile learning system which aims to support the semi-automatic adaptation of the learning activities. This is about the accommodations to: (a) the interconnection of the learning activities (i.e. the learning flow) and (b) the educational resources, tools and services that support the learning activities. The initial results of the assessment of the UoLmP use provide evidence that UoLmP can successfully be adapted to the learning flow of a pedagogical scenario and the provision of educational resources, tools and service that support the learning activities. This project includes three parts: capture/retrieval part, adaptation process part, and delivery/adjustment part. The Capture/retrieval part captures or senses the current situation properties for filtering learning contents, and detects current device capabilities for presenting the filtered contents polymorphically. The Adaptation process part executes the adaptation mechanisms, including the filtering mechanism and the polymorphic presentation mechanism, based on the IMS Learning Design Specification (IMS-LD). The Delivery/adjustment part delivers the adapted learning contents and learning activities to learners.

Some of them will, then, be analyzed in the mobile learning systems that have been performed to show the mobile learning features mentioned in the previous section.

5 Comparative Study

In this section, we give a comparative study of the different approaches and architectures of mobile learning that we have presented in the previous section.

This comparison is based on some significant criteria and features. We are particularly interested in:

1. *Device mobile support*: this is due to the fact that all the mobile devices used in the previous architectures are of personal/portable type (PDA, Smartphone, PC, Tablet, etc).
2. *Heterogeneity support:* the various hardware sensors, actuators, mobile devices with powerful servers, various network interfaces and different programming languages must be supported.
3. *Protection of privacy act*: flows of contextual information between system components must be controlled according to the needs and requirements of protection of users' privacy.
4. *Learning as a collaborative process,*
5. *The integration of formal and informal learning,*
6. *Learning as a set of activities in context:* it is to foster mobile technology in specific contexts to help carry out learning activities.
7. *Learning context-aware:* it is to help shift the monitoring of learning activities from one context to another in a mobile environment.
8. *Adaptability:* components that treated the context and communication protocols must adapt sufficiently in systems with a variable number of sensors, triggers and application components.
9. *Traceability and control:* the conditions of the system components and the flow of information between the components must be open to inspection so as to provide the users with adequate understanding and a control system.

10. *Tolerance at chess:* sensors or other components can possibly fail in the ordinary operation of a system. Disconnections may also arrive. The system should continue operations without demanding excessive resources, and detect failures.
11. *Deployment and configuration:* the hardware and software of the system must be easily deployed and configured to meet the requirements of users or environments, even for non-experts.

The result of this comparison is presented in Table 2. This table summarizes the capabilities of these architectures of context-aware systems. In fact, we find that none of the presented architectures fulfills all the criteria required for the implementation of a system sensitive to the context. The architectures layers (CAMLES and mCALS) and architecture of MoULe can be considered as models for system designers, but they still lack solutions to support privacy and tolerance failures. It is the same for architecture of eBag, which in addition does not support heterogeneity and traceability. The architecture of MOBIlearn has not context management. Learning as a set of activities in context is more important in the case of projects CMLearn and UoLmP in the other. Collaborative learning is most developed in the SAMCCO project in terms of participants' roles, pedagogical intention (collaborative missions) and supports. SAMCCO is based on the AM-LOM (Appliance Mastering LOM) metadata that is an extension of the LOM (Learning Object Metadata) metadata. AM-LOM the use of educational resources for indexing will allow several semantic ambiguity problems of some elements of the LOM and interpretation problems. For this, in our work we propose to use ontology for indexing educational resources based on LOM will allow a better understanding of the elements and securities offered and consequently facilitate their descriptions.

Most approaches do not enable the adaptation of the content to the learner's profiles. Context-aware approaches have the advantage of providing the user with the appropriate learning resources depending on the context. It is, therefore, essential to determine, depending on the context, how, when, and on which interface the resources

Table 2. Comparative study of the mobile learning architectures

Criterion	Approaches							
	MOBILearn	E-Bag	MoULe	mCALS	CAMLES	SAMCCO	CAMLearn	UoLmP
(1)	√	√	√	√	√	√	√	√
(2)	√	–	√	–	–	√	√	√
(3)	–	–	–	–	–	√	√	√
(4)	–	–	√	–	√	√	–	–
(5)	–	–	–	–	–	–	–	–
(6)	–	–	–	–	√	–	√	–
(7)	√	–	–	–	–	√	√	√
(8)	√	√	√	√	√	√	√	√
(9)	–	–	√	√	√	√	√	√
(10)	–	–	–	–	–	√	√	√
(11)	√	√	√	√	√	√	√	√

should be sent. However, Learning Through contextualization is not easy to achieve. In fact, the development of mobile technologies and the dynamics in the mobile environments have complicated the process of contextualization.

Today, mobile computing (user mobility, terminal mobility, and network mobility) is characterized by a permanent change in context (connected or disconnected, low or high bandwidth, change of location, widescreen or small screen, varying input devices, etc). Thus, it has become very complex to consider many and various aspects when designing such application.

It is worth-noting that mobile learning is very important, particularly in education, and the major utilization of mobile devices is in the field of medicine. In fact, medical students are placed in hospital/clinical environment require in their training an access to course information while on the move. In addition, the work of the postgraduates and the physicians involves a high degree of mobility between distributed sites and instant communications within work environments. Distributed sites, where physicians are working and in which students are placed, are often in remote and rural areas. The technological advances can be capitalized to promote and facilitate situated learning and collaborative.

In fact, because of the considerable growth of data, the heterogeneity of roles and needs as well as the rapid development of mobile systems, it becomes important to introduce a new system able to provide the users with a pertinent training adapted to their needs. We seek to develop an m-learning system of which the main issues are: (i) learning seen as "a collaborative process" that connects learners to communities of people through situations. Learners are not formed by a single teacher but by a learning community. (ii) Learning seen as "a process context-aware", the learning process must adapt to these changes in context: consideration of context aware and adaptation. This implies that the apprenticeship system is able to explore the environment to determine the current context and conduct learning activities in a particular context. Also, it is adapting learning resources (content, services, etc.) select the proper way to perform according to the current context activities.

In our work, a learning system named CCMLS (Context-aware and Collaborative Mobile Learning System) has been proposed. This system requires a context-aware architecture with mechanism that considers the change within the context. This architecture must take into account the users' different characteristics as well as all the contextual situations that influence their behavior when interacting with the mobile learning system. This system allows to share, to build, to collaborate with others remotely via collaborative tools (wiki, chat, forum, blog, etc.) or social networks of Universities, of Hospitals, etc.

The learner is only recipient of knowledge provided by the trainer but it becomes actor of the learning platform. He is involved in their own learning and working with the trainer and other learners. Finally, he shares his knowledge and expertise. We talk about learning community. Furthermore, the objectives of CCMLS can be specified as follows:

- To take not only learning objects, but also knowledgeable people as learning resources.
- The information about a role and individual role members should be used to help a learner to find appropriate knowledgeable people: such people may be a domain expert, teacher and even a co-learner when the learner needs to learn how to perform an activity.
- To integrate necessary communication tools.
- To classify the learning contents into learning objects, and to describe learning objects with metadata.
- To enable a learner to get the right learning objects at the right time.
- To enable a learner to contact the right people at the right time with the proper communication tool.
- To assist the learner to access the extensive knowledge artifacts' with which the learner could better understand the just learned knowledge or skills and also extend his learning area.

Through analysis of the relative work stated in previous section, three essential elements of a context-aware mobile learning system are: the context model, the learning units, and the adaptation engine with designed learning strategies. According to this structure, we state the overall architecture of the CCMLS system. This architecture consists of three layers: Learning Context Layer, Learning Adaptation Layer and Learning Application Layer. The context layer contains various physical and visual sensors to sense the learning context values defined in the context model. The goal of the Learning Adaptation Layer is to provide appropriate learning supports, including learning objects, learning community and learning activities, in relation with the current learning context. The learning application layer is in charge of interacting with learners, such as: collecting their information and requirements, displaying to them the adaptive learning objects, building the appropriate communication platforms between them and the selected learning community, helping them to complete proper learning activities, etc.

To realize and validate our proposals, a prototype of the system has been developed to facilitate the management of reusability and discovery of resources and services to deal with a dynamic environment. This prototype implemented with the PHP, Java, Jena, OWL, XML and MySQL technologies, using the development and running tools Eclipse, JDK, Protégé.

6   Conclusions

In this article, we have presented and compared the list of mobile learning architectures according to a set of features. All the presented architectures are not able to detect the mobile learning environment or to get information about it. Therefore, they cannot be adapted. Our work is an extension of one of the presented m-learning approaches in which we propose the context management, the adaptation and the learning without a break through contexts. Our goal is to design a mobile learning architecture that supports the features of mobility and context in order to enhance the learning experience in the field of education, specifically in medical field.

References

1. Sharples, M., Roschelle, J.: Guest editorial: special issue on mobile and ubiquitous technologies for learning. IEEE Trans. Learn. Technol. **3**(1), 4–5 (2010)
2. Sharma, S.K., Kitchens, F.L.: Web services architecture for m-learning. Electron. J. eLearn. **2**(1), 203–216 (2004)
3. Laouris, Y.: We need an educationally relevant definition of mobile learning. In: 4th World Conference on mLearning (MLearn), Cape Town, South Africa (2005)
4. Lam, J., Yau, J., Cheung, S.: A review of mobile learning in the mobile age. In: Tsang, P., Cheung, Simon, K.,S., Lee, Victor, S.,K., Huang, R. (eds.) ICHL 2010. LNCS, vol. 6248, pp. 306–315. Springer, Heidelberg (2010). doi:10.1007/978-3-642-14657-2_28
5. Sharples, M.: How can we address the conflicts between personal informal learning and traditional classroom education? In: Big Issues in Mobile Learning, Report of a Workshop by the Kaleidoscope Network of Excellence Mobile Learning Initiative, pp. 21–24. University of Nottingham (2006)
6. Lonsdale, P., Beale, R.: Towards a dynamic process model of context. In: Proceedings of UbiComp 2004 Workshop on Advanced Context Modeling, Reasoning and Manage (2004, to appear)
7. Brodersen, C., Christensen, B.G., Gronboek, K., Dindler, C., Sundararajah, B.: eBag: a ubiquitous web infrastructure for nomadic learning. In: Paper Presented at the Proceedings of the 14th International Conference on World Wide Web (2005)
8. Arrigo, M., Giuseppe, O.D., Fulantelli, G., Gentile, M., Novara, G., Seta, L., et al.: A collaborative m-learning environment. In: Paper Presented at the 6th Annual International Conference on Mobile Learning, Melbourne Australia (2007)
9. Yau, J.K., Joy, M.S.: A self-regulated learning approach: a mobile context-aware and adaptive learning schedule (mCALS) tool. Int. J. Interact. Mob. Technol. **2**(3), 52–57 (2008)
10. Nguyen, V.A., Pham, V.C., Ho, S.D.: A context - aware mobile learning adaptive system for supporting foreigner learning English. In: International Conference on Computing and Communication Technologies, Research, Innovation, and Vision for the Future (RIVF), IEEE RIVF, pp 1–6 (2010)
11. Yin, C.: SAMCCO: a system of contextual and collaborative mobile learning in the professional situations (in French), Thesis of Ecole Centrale de Lyon, Computer Science Department, 25 January 2010
12. SoualahAlila, F., Mendes, F., Nicolle, C.: A context-based adaptation in mobile learning. IEEE Comput. Soc. Tech. Committee Learn. Technol. (TCLT) **15**(4), p. 5 (2013)
13. Gómez, S., Zervas, P., Sampson, D., Fabregat, R.: Context-aware adaptive and personalized mobile learning delivery supported by UoLmP. J. King Saud Univ. Comput. Inf. Sci. **26**, 47–61 (2014)

Improved Document Feature Selection with Categorical Parameter for Text Classification

Fen Wang[1], Xiaoxuan Li[1(✉)], Xiaotao Huang[1], and Ling Kang[2]

[1] Department of Computer Science and Technology,
Huazhong University of Science and Technology, Hubei, China
pengpeng_1563@sina.com
[2] Department of Hydropower and Information Engineering,
Huazhong University of Science and Technology, Hubei, China

Abstract. Social network develops rapidly and thousands of new data appears on the Internet every day. Classification technology is the key to organize big data. Feature Selection (FS) is a direct way to improve classification efficiency. FS can reduce the size of the feature subset and ensure classification accuracy based on features' score, which is calculated by FS methods. Most previous studies of FS emphasized on precision while time-efficiency was commonly ignored. In our study, we proposed a method named CDFDC at first. It combines both CDF and Category-Frequency. Secondly, we compared DF, CDF, CHI, IG, CDFP_VM and CDFDC to figure out the relationships among algorithm complexity, time efficiency and classification accuracy. The experiment is implemented with 20-newsgroup data set and NB classifier. The performance of the FS methods evaluated by seven aspects: precision, Micro F1, Macro F1, feature-selection-time, documents-conversion-time, training-time and classification-time. The result shows that the proposed method performs well on efficiency and accuracy when the size of feature subset is greater than 3,000. And it is also discovered that FS algorithm's complexity is unrelated to accuracy but complexity can ensure time stability and predictability.

Keywords: Feature selection · Measurement · Comparison · Time efficiency · Experimentation

1 Introduction

With the popularity of social network, the amount of data that's being stored on the Internet is almost inconceivable and it just keeps growing. Information retrieval becomes more and more difficult without classification technology [1]. Classification is the key to classify vast amount of data [2]. It has been applied to many kinds of fields such as text classification, topic detection, spam filtering, web-page classification, sentiment classification [3], web-link classification, biomedical information classification, engineering test data classification etc.

As the basis of all classification problems, text classification is still an important aspect to study. There are two ways of text representation: the original text

© Springer International Publishing AG 2016
S. Boumerdassi et al. (Eds.): MSPN 2016, LNCS 10026, pp. 86–98, 2016.
DOI: 10.1007/978-3-319-50463-6_8

and the Vector Space Model (VSM). And VSM is proved a more efficient method by previous studies [4].

But in text categorization problem, VSMs dimension is always high because vocabulary scale of one language is always huge. The high dimensional vector could bring the curse of dimensionality [2,5]. Therefore we usually reduce dimension of a VSM to improve time-efficiency and accuracy of text classification, which can be implemented by feature extraction or feature selection methods [6]. In [7,8], it has been proved that a good feature selection method has positive effect on building robust machine systems.

During our research work, we compare five typical feature selection methods: Document Frequency (DF), Categorical Document Frequency (CDF), CHI, IG and CDFP_VM [2]. And we evaluate them by seven criteria: precision, Micro F1, Macro F1, feature selection time, document-conversion-time, training-time and testing-time. The results show that algorithm's complexity is unrelated to classification accuracy but has positive effect on time stability and predictability. Besides, we proposed a new feature selection method Categorical Document Frequency Divided by Categorical Number (CDFDC) to improve CDF. Its advantages are:

- Low computational complexity and improved classification accuracy.
- Shorter and more stable on classification time.

The paper is organized as following. Section 2 overviews and summarizes feature selection methods. Section 3 briefly describes the common feature selection methods. The proposed feature selection method is described in Sect. 4. The experimental results and analysis are shown on Sect. 5. Section 6 draws conclusions and outlines the future work.

2   Feature Selection Methods Overview

As a key to improving the efficiency and accuracy of text classification, feature selection method has been the research focus all the time. We divide previous feature selection methods into five kinds.

The first kind: feature selection methods based on statistics. Such methods focus attention on term frequency [2,5,9]. CHI and IG are proved mostly to have better classification results. But with slightly less accuracy, among all the methods, DF and CDF are the best. CDFP_VM [2] stresses the distribution of class by computing variance to extract features. FEDIP [5] calculates correlation using conditional probability. Both of them produce low-dimensional feature space while maintaining the accuracy of classification. The obvious defect is that the results are not universal and convincing enough because the experimental-data-sets' size is smaller than 3,000.

The second kind: vector-distance-based feature selection method. These methods select the most relevant features by calculating correlation coefficient or vector distance [10]. DET [6] calculates the average distance of feature and classes under a condition and many conditions to obtain the DE factor. Research work proved that the similariy between features within the class and class discrimination are significantly influenced by the DE factors.

The third kind: the optimization algorithm based feature selection method. Common algorithms include PCA, factor analysis, feature scaling, PSO [6,11], decision trees, etc. These methods are commonly able to obtain a high classification accuracy, along with the high computational complexity.

The fourth kind: pattern-based feature selection method. [12] proposed RFD feature selection method, which will discover positive and negative modes of advanced features, and deploy both modes to the low-level features. With this method, the synonym and polysemy problems are solved.

The fifth kind: the combination of feature selection methods. [6,8] proposed two different Two-stage feature selection methods. They used one method to select features and then applied another method to reduce features' vector space dimension again. The combinations can get smaller vector space and the selected features tend to be more correlative with categories. For different applications [9,11–13], combinations are different.

In summary, the first kind of method has low computational complexity and wide application. But other types are more suitable to problems with smaller data sets or higher precision request. And their computational complexity is usually high. Our novel method belongs to the first kind.

3   Conventional Methods

3.1   Document Frequency (DF)

For each term t, DF counts the number of documents including t every class respectively and the maximum one will be t's score. This is the most basic idea of all feature selection methods. However, DF has no consideration to the feature-class distribution.

3.2   Categorical Document Frequency (CDF)

CDF is an improved version of the DF. Let there be $\{c_j : j = 1, 2, ..., n\}$ classes in target space. CDF counts the number of documents in c_j class for each item to reflect the feature-class distribution.

3.3 χ^2 Statics (CHI)

CHI calculates the correlation coefficient between the term and class. CHI's formula is defined as

$$\chi_{t,c}^2 = \frac{(P(t,c)P(\bar{t},\bar{c}) - P(t,\bar{c}))P(\bar{t},t))^2}{P(t)P(\bar{t})P(c)P(\bar{c})} \tag{1}$$

$$\chi^2(t) = \sum_{j=1}^{n} P(c_j)\chi_{t,c}^2 \tag{2}$$

3.4 Information Gain (IG)

IG uses the conditional probability to calculate correlation coeffecient between term t and class c_j. And the *log* operation can reduce the loss of precision. IG is defined below.

$$IG(t) = -\sum_{j=1}^{n} P(c_j)log(P(c_j) + P(t)\sum_{j=1}^{n} P(c_j|t)log(c_j|t) + P(\bar{t})\sum_{j=1}^{n} P(c_j|\bar{t})logP(c_j|\bar{t})$$

(3)

3.5 CDFP_VM

CDFP_VM is a little complicated. $CDFP(t, c_j)$ gets the proportion of term t based on the CDF by divided the number of documents in c_j class.

$$CDFP(t, c_j) = CDFP(t, c_j)/|c_j|$$

(4)

Obtain $\{CDFP(t, c_j) : j = 1, 2, ..., n\}$ expectations

$$w_t = \{CDFP(t, c_1), ..., CDFP(t, c_n)\}$$

(5)

$$\overline{w_t} = \frac{\sum_{j=1}^{n} CDFP(t, c_j)}{n}$$

(6)

Calculate variance of term t. According to the definition in statistics, great variance means term t has the dicrimination among classes. Meanwhile when the variance is small enough, term t can be removed as same as stop words. CDFP_VM(t) is defined as

$$CDFP\_VM(t) = \frac{\sum_{j=1}^{n}(CDFP(t, c_j) - \overline{w_t})^2}{n}$$

(7)

4 Improved Method CDFDC

4.1 CDF's Weak Point

CDF is widely used in text classification. But its calculation method makes some high-score features likely occur in many classes, which means that these selected features' value is as low as stop words. Let us see one example, where REC is a class including four small classes: AUTOS, MOTOCYCLES, BASEBALL and HOCKY. We can see the documents in Table 1.

After removing stop-words and stemming of the document set, we can get terms' statistical result as shown in Table 2.

The word scores calculated by the CDF method is shown as:

$$CDF(rec) = CDF(autos) = CDF(baseball) = CDF(motocycles) = 2 >$$
$$CDF(sport) = 1.$$

Table 1. Example's document set

DocNo	Category	Content
1	AUTOS	This is rec autos
2	MOTOCYCLES	This is rec motocycles
3	BASEBALL	This is baseball and it's belonging to sport
4	BASEBALL	Baseball is belonging to rec
5	AUTOS	Autos is rec
6	MOTOCYCLES	I am watching motocycles, which is a sport

Table 2. Terms statstics

	AUTOS	MOTOCYCLES	BASEBALL
rec	2	1	1
motocycles	0	2	0
sport	0	1	1
baseball	0	0	2
autos	2	0	0

It is clear that term *rec* appears in almost all documents like stop words. Since term *rec* has little value in classification work, it should be removed. But according to CDF rule, it gives score of term *rec* as equivalent as it of term *autos*, *baseball* and *motocycles*. Besides, the score of term *rec* is even higher than term *sport*. It is obviously unreasonable. This situation is caused by that different text classification problems always have different stop-word lists. A term is not that important if it appears in many categories of a document set. This is a common perception. To revise this weak point, we define the number of categories including term t as CF(t) and add CF(t) in the formula of CDF to increase terms' discrimination. The proposed improved feature selection based on CDF is named as CDFDC (Categorical Document Frequency Divided by Categorical Number).

4.2 CDFDC Method

CDF's method is shown below.

$$CDF(t) = Max\{CDF(t,c_1),...,CDF(t,c_n)\} \tag{8}$$

c_j: Category NO is j and $j = \{1,2,...,n\}$
$CDF(t,c_j)$: Number of documents which include term t

Firstly, we need to know whether term t appears in the class c_j or not, which defined as $CF(t,c_j)$

$$CF(t,c_j) = \begin{cases} 0 & ,t \in c_j \\ 1 & ,t \notin c_j \end{cases}$$

Secondly, CF(t) summarizes the frequency of term t appears in categories set $\{c_1, ..., c_j, ..., c_n\}$, which is called for short as CF (Category Frequency) and it is defined as:

$$CF(t) = \sum_{j=1}^{n} CF(t, c_j) \tag{9}$$

We divide $CDF(t)$ by $CF(t)$ to reduce the value if a term t appears in more than one category. The method can increase feature's class discrimination. And we give $CDFDC(t)$ definition following.

$$CDFDC(t) = \frac{CDF(t)}{CF(t)}. \tag{10}$$

By definition CDFDC, we recalculate feature scores of the example and obtain the following result:

$CDFDC(autos) = CDFDC(baseball) = CDFDC(motocycles) = 2/1 = 2 > CDFDC(rec) = 2/3 = 0.67 > CDFDC(sport) = 1/2 = 0.5$.

Obviously, in this example, scores calculated by CDFDC are more reasonable than CDF. In Sect. 5, we'll do experiments to prove our vision.

5   Experiments and Analysis

This section shows the performance of six feature selection methods: DF, CDF, CHI, IG, CDFP_VM and CDFDC. NB classifier is chosen since it performs very well in accuracy and efficiency of text classification [1]. The VSM representation of documents used for classification is given below. And there are seven evaluation criteria: precision, Micro-F1, Macro-F1, feature selection time, document conversion time, training time and test time. Dataset used in comparison is 20-news-group and the tests choose the number of features from 1,000 to 13,000.

5.1   Text Preprocessing Step

Text pre-processing consists of three steps: word segmentation, removing stop words and stemming. We use Lucene Pack *CJKAnalyzer* to implement word segmentation. The whole project is based on Java. After that, original dictionary can be generated. Then, feature selection methods are used to score terms in original dictionary and extract most valuable ones building feature set or feature dictionary. In our experiment, we convert the documents into sparse matrix, which also can be called VSM. VSM is proved to be a more efficient way in text classification. The format of the VSM we used is defined below:

label term-no: term-document-frequency

For example, a document belonging to class 1 and its content is "Hello, world", in which *hello*'s term-no is 23 and *world*'s term-no is 100. The document is converted to vector as.

1 23:1 100:1

In order to improve time efficiency and facilitate the dynamic growth of document set in the future, we convert the training set and testing set into two text files to store VSMs: train.matrix and test.matrix. Finally, NB classifier is used to execute classification mission meanwhile the evaluation criteria are calculated or recorded.

5.2   Data Set Description

In this paper, the data set is 20-news-group. It is a standard data set commonly used in the English classification problem and it contains 20 classes, each containing approximately 1,000 documents. The documents set is divided into two parts: training documents set called 20news-bydate-train and test documents set called 20news-bydate-test. The training set contains 20 classes and 11,315 txt files; testing set contains 20 classes and 7,533 txt files. Distributed of the two documents sets are shown in Table 3.

Table 3. 20-news-group documents distribution

Training set name	ClassNO	FilesNum	Testing set name	ClassNO	FileNum
20news-bydate-train	1	480	20news-bydate-test	1	319
	2	584		2	389
	3	591		3	394
	4	590		4	392
	5	578		5	385
	6	593		6	395
	7	585		7	390
	8	594		8	396
	9	598		9	398
	10	597		10	397
	11	600		11	399
	12	595		12	396
	13	591		13	393
	14	594		14	396
	15	593		15	394
	16	599		16	398
	17	546		17	364
	18	564		18	376
	19	465		19	310
	20	377		20	351

5.3   Evaluation Criteria

Evaluation parameters Micro-F1 and Macro-F1 is commonly used text classification evaluation criteria. To class c_i, which be within $\{c_i : i = 1, 2, ..., n\}$ for n classes, here are the relevant formulars calculating accuracy, recall rate and F1 standards.

$$P_{c_i} = \frac{TP_{c_i}}{TP_{c_i} + FP_{c_i}} \tag{11}$$

$$R_{c_i} = \frac{TP_{c_i}}{TP_{c_i} + FN_{c_i}} \tag{12}$$

$$F_{c_i} = \frac{2P_{c_i}R_{c_i}}{P_{c_i} + R_{c_i}} \tag{13}$$

TP_{c_i}: Number of files assigning to c_i correctly
FP_{c_i}: Number of files assigning to c_i wrongly
FN_{c_i}: Number of files which are not assigning to c_i but actually belonging to c_i

We can calculate the average of Micro F1 and Macro F1 to measure the overall performance. Macro-F1 is defined as

$$F(Macro - average) = \frac{\sum_{i=1}^{n} F_{c_i}}{n} \tag{14}$$

Micro-F1 is defined as

$$P_{micro} = \frac{\sum_{i=1}^{n} TP_i}{\sum_{i=1}^{n} (TP_i + FP_i)} \tag{15}$$

$$R_{micro} = \frac{\sum_{i=1}^{n} TP_i}{\sum_{i=1}^{n} (TP_i + FN_i)} \tag{16}$$

$$F(Micro - average) = \frac{2P_{micro}R_{micro}}{P_{micro} + R_{micro}} \tag{17}$$

5.4   Experiment Results and Analysis

Experiment is to conduct text classification using dataset 20-news-group with NB classifier based on VSM. The contrast feature selection algorithms include DF, CDF, CHI, IG, CDFP_VM and CDFDC; comparative classification measures are precision, Micro F1, Macro F1, feature selection time, document conversion time, training time and testing time, which are shown in Figs. 1, 2, 3, 4, 5, 6 and 7 respectively. The abscissa is the first n feature points, from 1,000 to 13,000 and ordinate indicates accuracy (decimal) or time (seconds).

Figure 1 shows that CDFDC, CHI, DF precisions are less than IG, CDF and CDFP_VM at beginning, in which IG performs best on 0.7080; CHI performs worst on 0.6260. But CDFDC performance starts to outperform other feature selection methods when the number of features is bigger than 3,000. CDF and

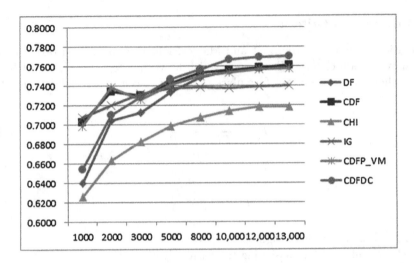

Fig. 1. Classify precision

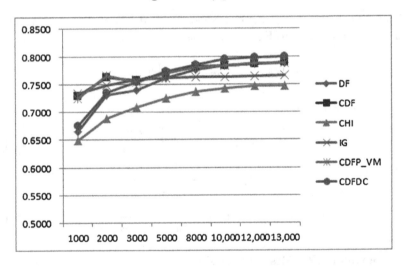

Fig. 2. Micro-F1

CDFP_VM precision lines coincide substantially. Surprisingly, CHI precision performance is always the worst.

Figures 2 and 3 are Micro F1 and Macro F1 line charts, 3,000 is still the number of cut-off point. After 3,000, CDFDC performance tends to be good. Wherein IG performance is the most stable, and the performance of CDFP_VM method and CDF method are always consistent.

Figure 4 shows the time of feature selection. As it shows, the time of CDFDC, DF and CDF are shorter. IG's time is longest, which is because IG has the highest algorithm complexity. Meanwhile, the time line of two algorithms: CDFP_VM and CHI are almost duplicate, which is to coincide with the similarity of their complexity.

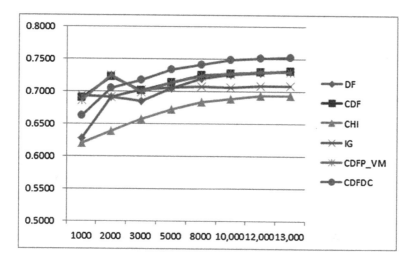

Fig. 3. Macro-F1

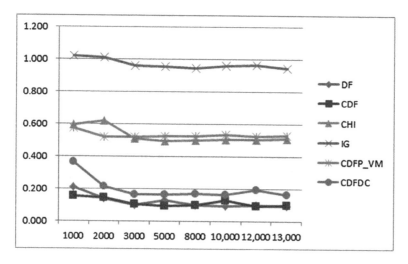

Fig. 4. Feature selection time

Figure 5 shows the time converting training set and testing set into sparse matrixes. CDF time performance is extremely unstable. It indicates that feature set selected by CDF does not appear evenly in each file. CDFDC and CDFP_VM performance is relatively stable. The spending-times are in accordance with the size of selected feature set, which is regular. The stable-feature-selection-time-lines suggest that CDFDC and CDFP_VM extract feature sets which are most consistent with files.

Figures 6 and 7 show the training time and testing time respectively. We can see CDFP_VM polyline fluctuate greatly, which has basically nothing to do with

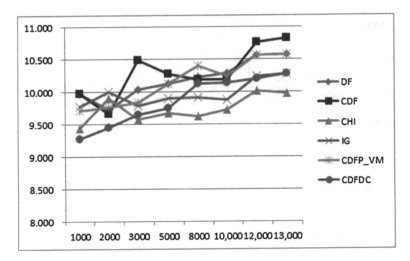

Fig. 5. Document converting time

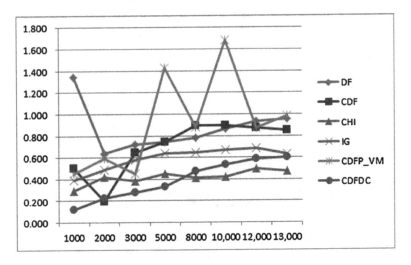

Fig. 6. Training time

the number of features. CDF training time fairly stable and predictable, but it has greatly fluctuated when the number of selected features is smaller than 5,000. DF's training time reaches up to 1.347 s when there are 1,000 extracted features, which is unexpected. DF's classification times have nothing to do with features totally. IG's time lines are relatively flat, but they are always a little longer than CHI and CDFDC. CDFDC always performs best while the feature set size is under 8,000 and it is still the top 2 method after that.

All in all, we can sum up two conclusions from the Figs. 1, 2, 3, 4, 5, 6 and 7. Firstly, as for DF, CDF, CHI, IG and CDFP_VM, the algorithm complexity

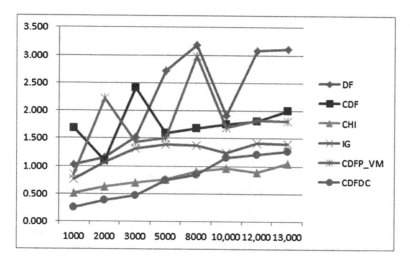

Fig. 7. Testing time

has no direct relationship with time efficiency, precision and recall. CDF has much lower algorithm complexity than CDFP_VM, but polylines of CDF and CDFP_VM always coincide substantially in Figs. 1, 2, 3, 4, 5, 6 and 7. In other word, high complexity may not bring better classification results, although you can get a more stable and predictable time efficiency. Secondly, CDFDC performs superior to the other in precision, Micro F1 and Macro F1, when the number of selected features is greater than or equal to 3,000. Even not best, CDFDC also show relatively low and predictable stability time efficiency. The novel method CDFDC has a good overall performance.

6   Conclusion

This paper does two aspects: firstly, we implement a comparison among 4 common methods, CDFP_VM which is proposed in 2014 and CDFDC on classification efficiency; secondly, we proposed a new method called CDFDC combining both CDF method and Catogory Frequency. We found that CDFP_VM doesn't perform significantly better than the other feature selection methods in English date set, especially not better than CDF. This shows that high computation complexity may not bring high precision or recall rate, except that we need stable and predictable time efficiency. Totally, we can draw CDFDC as a good feature selection method after summing up CDFDC's performs in precision and time efficiency comparing to other methods. But in this paper, the data set is not large enough and not so much practical. The future work is to experiment on larger data set and find the connection between algorithm complexity and documents-time-efficiency.

Acknowledgments. I feel much indebted to many people who have instructed me in writing this paper. I would like to express my heartfelt gratitude to my tutor, Prof. Wang, for her warm-heart encouragement and most valuable advice, especially for her insightful comments and suggestions on the draft of this paper. Without her help, encouragement and guidance, I could not have completed this paper.

And I would like to express my thanks to my family and my friends for their valuable encouragement and spiritual support during my study.

References

1. Basu, T., Murthy, C.A.: Effective text classification by a supervised feature selection approach. In: 12th IEEE International Conference Data Mining Workshops (ICDMW), pp. 918–925. IEEE Press, New York (2012)
2. Li, Q., He, L., Lin, X.: Improved categorical distribution difference feature selection for Chinese document categorization. In: 8th International Conference on Ubiquitous Information Management and Communication, pp. 102:1–102:7. IEEE Press, New York (2014)
3. Sharma, A., Dey, S.: A comparative study of feature selection and machine learning techniques for sentiment analysis. In: ACM Research in Applied Computation Symposium, pp. 1–7. IEEE Press, New York (2012)
4. Wang, Z., Chen, S., Liu, J., Zhang, D.: Pattern representation in feature extraction and classifier design: matrix versus vector. J. IEEE Trans. Neural Netw. **19**, 758–769 (2008)
5. Tariq, A., Karim, A.: Fast supervised feature extraction by term discrimination information pooling. In: 20th ACM International Conference on Information and Knowledge Management, pp. 2233–2236. IEEE Press, New York (2011)
6. Van, M., Kang, H.-J.: Bearing-fault diagnosis using non-local means algorithm and empirical mode decomposition-based feature extraction and two-stage feature selection. J. IET Sci. Measur. Technol. **9**, 671–680 (2015)
7. Somol, P., Novovicova, J.: Evaluating stability and comparing output of feature selectors that optimize feature subset cardinality. J. IEEE Trans. Pattern Anal. Mach. Intell. **32**, 1921–1939 (2010)
8. Meng, J., Lin, H.: A two-stage feature selection method for text categorization. In: Seventh International Conference Fuzzy Systems and Knowledge Discovery (FSKD), pp. 1492–1496. IEEE Press, New York (2010)
9. Zhang, W., Yoshida, T., Tang, X.: A comparative study of TF*IDF, LSI and multi-words for text classification. J. Exp. Syst. Appl. **38**, 1492–1496 (2011)
10. Kadhim, A.I., Cheah, Y.N., Ahamed, N.H., Salman, L.A.: Feature extraction for co-occurrence-based cosine similarity score of text documents. In: IEEE Student Conference Research and Development (SCOReD), pp. 1–4. IEEE Press, New York (2014)
11. Li, Y., Algarni, A., Albathan, M., Shen, Y., Bijaksana, M.A.: Relevance feature discovery for text mining. In: IEEE Transactions on Knowledge and Data Engineering, pp. 1656–1669. IEEE Press, New York (2015)
12. Li, Y., Algarni, A., Albathan, M., Shen, Y., Bijaksana, M.A.: Relevance feature discovery for text mining. J. IEEE Trans. Knowl. Data Eng. **27**, 1656–1669 (2015)
13. Song, S.J., Heo, G.E., Kim, H.J., Jung, H.J., Kim, Y.H., Song, M.: Grounded feature selection for biomedical relation extraction by the combinative approach. In: ACM 8th International Workshop on Data and Text Mining in Bioinformatics, pp. 29–32. IEEE Press, New York (2014)

A Geographic Multipath Routing Protocol for Wireless Multimedia Sensor Networks

Mohamed Nacer Bouatit[1]([✉]), Selma Boumerdassi[1], and Adel Djama[2]

[1] CNAM/CEDRIC, Paris, France
mohamednacer.bouatit@auditeur.cnam.fr, selma.boumerdassi@cnam.fr
[2] ESI, Algiers, Algeria
a_djemaa@esi.dz

Abstract. The availability of low-cost hardware such as cameras and microphones has fostered the development of wirelessly interconnected devices that are able to retrieve from the environment, in addition to scalar data, the audio and video streams. Transfering the large multimedia content while guaranteeing QoS requirements is a very challenge to solve, given that multimedia applications convey a large mass of data that requires high transmission rate and intensive treatment. Therefore, in this paper, we propose a Geographic Multipath Routing Protocol (GMRP) that satisfies the realistic characteristics of multimedia transmission in wireless sensor networks.

Simulations results show a significant contribution and indicate that GMRP achieves lower average delay, higher packet delivery ratio and ensures load balancing in the network.

Keywords: Wireless multimedia sensor networks · Geographic routing protocol · Multipath transmission · Load balancing

1 Introduction

Over the past few years, progress in Complementary Metal-Oxide Semiconductor (CMOS) technology has facilitated the development of low-cost, physically small smart sensors to sense multimedia data [1]. This development gave birth to a new type of network called Wireless Multimedia Sensor Network (WMSN), which allows to enrich existing applications using WSNs, and brings a new level of innovative applications which provide better description of captured event and enhance quality of collected information.

WMSNs can contain multiple types of heterogeneous sensors (scalar or multimedia) with low or high resolution, cooperating with each other in interest area. Data collected by the sensors are sent to collector to extract useful information. This task is entrusted to the routing protocols in charge of routing the traffic end-to-end using a multi-hop strategy. In addition to the inherited constraints of classical sensors networks (energy, storage and computing), routing protocols in WMSNs face several challenges, particularly to ensure the QoS requirements and

© Springer International Publishing AG 2016
S. Boumerdassi et al. (Eds.): MSPN 2016, LNCS 10026, pp. 99–108, 2016.
DOI: 10.1007/978-3-319-50463-6_9

to better manage the limited energy of sensors nodes, which consequently reduces network lifetime and causes instability during data transmission [2]. Moreover, the introduction of multimedia data and the growing interest in real-time applications have made strict constraints on throughput and delay to better transmit data at critical time and without any loss.

In this paper, we propose a routing protocol that takes into account the following requirements according to multimedia transmission:

1. *Multipath transmission*: the large size of multimedia flow and transmission requirements can exceed sensor's capacity to support the whole stream.
2. *Shortest path node-disjoint*: multimedia applications have time constraints, which requires the use of the shortest path node-disjoint with minimal end-to-end delay.
3. *Hole-bypassing*: the bypass of static holes is necessary to guarantee the reliability.

The remainder of this paper is organized as follows: Sect. 2, discusses routing protocols proposed to multimedia transmission and gives a short classification of existing solutions. Section 3, presents network model. Proposed Geographic Multipath Routing Protocol is described in Sect. 4. Results of extensive simulations are shown in Sect. 4. Finally, Sect. 5, concludes the paper.

2   Related Work

Several routing protocols have been proposed for WSNs, however, they are not suitable for WMSNs due to the new features and constraints imposed by the intrinsic nature of multimedia stream [3–5]. The recent research in WMSNs aims at meeting the QoS requirements by applying modifications into existing protocols or proposing new mechanisms and strategies.

Our reading routing solutions found in literature, has led us to highlight the following main classes, basing essentially on design principle and operation mode of protocols, namely:

– Protocols based on *Multipath/Single Path transmission*: use a preventive algorithm to establish more then one path from the source node to the sink, in order to increase transmission performance, ensure the reliability and load balancing TPGF [6], GEAM [7] and ADRS [8]. This offers a rapid recovery of transfer in case of failures, unlike protocols that use a single routing path as McTPGF [9], MREEP [10] and PASPEED [11].
– Protocols based on *Meta-heuristics*: they appeal to methods in the field of operational research for making the best routing decision. These techniques are inspired by the nature, such as the optimization based on ant colonies AntSensNet [12] and QoS-LB [13]. These solutions overload the network with different control packets which increase power consumption and bandwidth use. Also, an addtional cost of computing time and memory caused by crossover and mutation operations of chromosomes for paths evaluation.

- Protocols based on *Cross-layer design*: to ensure the different QoS parameters in each layer with the guarantee of interdependence between these layers. These protocols have led the designers to rely on a cross-layer solution, which aims to optimize the performance of the routing and to provide a delivery guarantee with low-resource consumption, like A Cross-Layer-Based Clustered Multipath Routing with QoS-Aware Scheduling for WMSN [14], MEVI [15] and EEQR [16]. These solutions do not address all QoS parameters, in addition to the use of complex methods for packet scheduling and allocation of communication channels.
- Protocols based on *Geographic location*: use the geographic coordinates of sensor nodes for routing data. From this class, two routing variants have been derived:
  - The offline routing is done in two steps, path construction and sending data, AAEEGF [17], MPMPS [18] and EA-TPGF [19],
  - The online routing is done gradually when creating paths (hop by hop). AGEM [20], MMSPEED [21].

In this article, we focused on geographic routing protocols because of their effectiveness in optimal path discovery and we have also used the multipath transmission in order to ensure QoS requirements.

3   Network Model

We consider a flat network architecture and the wireless sensor network is composed of N sensors, deployed in a static deterministic manner, each sensor node being aware of its geographic location and its 1-hop neighbor nodes? geographic locations. We assume that only source nodes knows Base Station (BS) location and all other sensor nodes know BS location by receiving packets from source nodes. All nodes have the same transmission range and are homogeneous and are endowed with identical physical capabilities (detection and communication). Only bidirectional links are used to build paths. Each sensor node may be in one of the following states:

- Valid: ready to build a path;
- Active: already used in a path (locked for specific path);
- Blocked: no valid next-hop except its predecessor;
- Failed: broken, damaged or exhausted battery.

Static holes are the subset of failed or blocked nodes and each path is composed of a finite set of links, each node can belong to only one path, except source nodes and sink node. In GMRP, all generated routing paths are node-disjoint routing paths, this assumption is the same with that used in [2,6,9,17,19].

4   Proposed Protocol

Motivated by satisfying multimedia transmission requirements, GMRP consists of two phases: Geographic forwarding and Path optimization. It takes place after the deployment of the sensors and from the periodical exchange of beacon messages, it filled all the neighborhoods tables of nodes in the network.

4.1 Messages Used

In our design, we have use three types of messages, namely: beacon message, control message during paths exploration phase and data message to transfer multimedia stream.

1. *Beacon message:* is broadcast periodically by each node to communicate with its neighbors the useful information for making the routing decision, such as the ID, the coordinates (X, Y) and the current state of the node. Neighborhoods tables are updated upon beacon messages reception, but also following the receipt of control messages during path's exploration.
2. *Control messages:* exchanged through paths exploration phase, they are the basis of GMRP operation (Table 1). Their format includes five categories:
   - *GreedyForwarding*: to explore the path.
   - *AckGreedyFowarding*: to carry the GreedyForwarding message.
   - *Walking Back*: to return to the predecessor in case of blocking situation.
   - *Optimization*: this message is sent to the nearest predecessor from the source node to optimize the found path.
   - *Release*: sent in two cases, after receiving the optimization message to release the unoptimized nodes, or at the end of data transmission, in order to release nodes belonging to the paths.
3. *Data message:* once maximum path number reached (at least one found path), the protocol begin transferring data. Data message format is given in Table 2.
   PacketSN: packet sequence number.
   PacketTime: cumulative time of packet from source node through the intermediate nodes until arrival to sink node.
   Data: useful part of data captured by source node.

Table 1. Control message format.

MsgType	NodeState	SourceID	SinkID	PathID	Label
1 byte	1 byte	2 bytes	2 bytes	2 bytes	2 bytes

Table 2. Data message format.

PacketSN	PacketTime	Data
2 bytes	4 bytes	1..N bytes

4.2 Operating Phases

GMRP proceeds after neighborhood acknowledgment in two phases: path exploration through geographic routing and optimization of found path by removing loops.

Path Exploration: when an event occurs, source node initialize the field *Label* to the maximum value, registers path signature (SourceID, SinkID and PathID), changes its status to active and launches paths exploratorion phase by sending a *GreedyForwarding* message to the nearest neighbor from the Base Station (BS). If it exists, it marks the node as its successor, changes its status to active in its neighborhood table.

This neighbor sends *AckGreedyFowarding* message to its predecessor to inform him of its state: is-it active or not? Indeed, *GreedyForwarding* request would be taken over by the neighbor only if it does not participate in any other path, given that routing paths are node-disjoint routing paths.

If the issuer learns, via the acknowledgment message that its neighbor is already active, it starts another path exploration phase. In case of a favorable response from the neighbor, this last updates its neighborhood table, saves path signature, decrements the field *Lebel* and explores the routing path in the same way as its predecessor until reaching the BS. Source node repeats this process with all of its neighbors until achieving the maximum path number definite.

In case of blocking of explorer node (e.g., it is unable to find a next-hop node), it rolls back to its predecessor that marks it as a blocked node, and restarts another query exploration path. Once sink node receives the first packet, it starts optimization phase using reverse shortest path of exploration until reaching source node. The flowchart of Fig. 1(a) summarizes path exploration phase.

Path Optimization: if a node receives a query optimization but not from its direct successor of the previous phase (exploration), it concludes that an optimization has occurred, and releases its successor. The aim of this phase is to release

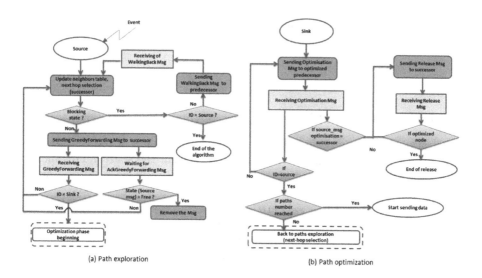

(a) Path exploration   (b) Path optimization

Fig. 1. Operating phases flowcharts.

nodes, to participate in construction of other paths and therefore, increase network availability in terms of free nodes. Figure 1(b) shows optimization phase flowchart.

Once both exploration and optimization phases have been completed, source node starts sending data packets by sharing flow between different built paths.

5 Evaluation

We present and discuss in this section our simulation results for the performance study of our GMRP. We used TinyOs [22] environment which relies on low power consumption operations to implement and conduct a set of simulation experiments for our protocol and compare it with MREEP and TPGF protocols. MREEP based on distance and hop count metrics and supporting both real-time and non real-time traffic using single path transmission while TPGF protocol, which is a geographic routing protocol cited in several recent papers, uses multipath transmission with an efficiency in optimal path discovery (Table 3).

Table 3. Main configuration parameters.

Parameter	Value
Network size	500 m × 250 m
Sensors number	50–300
Bandwidth	250 kbits/s
Transmission range	22 m
Packet size	128 bytes
Data size	6.4 MB
Path number	5
Queue size	30

5.1 Delivery Ratio (DR)

Is the ratio, for a given period of time, between the number of received packets N_{RP} and the total number of sent packets N_{SP} from the source. This metric reflects the reliability of the protocol during packets? shipping from the source to the destination.

$$DR = \frac{N_{RP}}{N_{SP}} \qquad (1)$$

One can note from Fig. 2 that the delivery ratio is proportional to paths number used for both protocols. Indeed, when the number of paths is reduced, sensor nodes are more loaded and queues are faster overflowed and a more important packet loss is recorded. From this, one can infer that for broadband, the larger the number of paths for the routing protocol, the better the delivery rate.

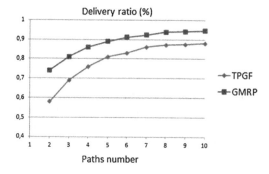

Fig. 2. Impact of paths number on delivery ratio.

5.2 Average End-to-end Delay (AED)

Is the sojourn time of the packets in the network from their issue by the source, until their arrival at the collector (this time includes the processing time, the residence times in the waiting queues, the transmission time being ignored). If packet j passes through a path containing n intermediate nodes, the end-to-end average delay is given by:

$$AED_i = \sum_{k=1}^{n} |T_{in}^k - T_{out}^k| \tag{2}$$

where T_{in}^k is the arrival time of packets on node k and T_{out}^k is the output time of the packet from the waiting queue for this node. Thus, the average end-to-end delay of all packets:

$$AED = \frac{1}{N_{RP}} \sum_{j=1}^{N_{RP}} EAD_i \tag{3}$$

Figure 3 highlights that the average end-to-end delay in GMRP is shorter than the one of MREEP. The residence time of the packets in the queue is

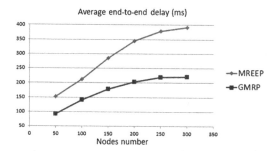

Fig. 3. Impact of nodes number on average end-to-end delay.

inversely proportional to operational paths number. The increase of paths alleviates queues and reduces congestion.

5.3   Energy Distribution (ED)

Is the energy distribution in the network, this metric allows to ensure a load balancing and it is calculated by the variance of residual energy of all network nodes, where N represents the number of sensors constituting the network and E_k is the energy consumed by node k.

$$ED = \frac{1}{N} \sum_{k=1}^{N} (E_k - E_{avg})^2 \tag{4}$$

N: Total number of node.
E_k: Residual energy of node k.
E_{avg}: Average residual energies of all nodes.

The energy consumption is lower with GMRP compared to MREEP, as illustrated in Fig. 4. Multipath transmission alleviates nodes and shares multimedia stream between different found paths, which consequently allows to have a load balancing in the network.

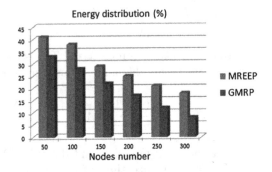

Fig. 4. Impact of nodes number on energy distribution.

6   Conclusion

In this work we presented a geographic routing protocol dedicated to multimedia stream, which takes into account multipath transmission, shortest path node-disjoint and bypasses holes. The obtained results prove that GMRP provides best performance and meets the QoS requirements for multimedia transmission, especially in terms of delivery rate and average end-to-end delay. Moreover, it also provides load balancing by distributing energy between operational paths during transmission.

References

1. Aswale, S., Ghorpade, V.R.: Survey of QoS routing protocols in wireless multimedia sensor networks. J. Comput. Netw. Commun. **2015**, 1–29 (2015)
2. Bouatit, M.N., Boumerdassi, S., Minet, P., Djama, A.: Fault-tolerant mechanism for multimedia transmission in wireless sensor networks. In: Vehicular Technology Conference VTC2016-Fall, p. 84 (2016)
3. Pantazis, N.A., Nikolidakis, S.A., Vergados, D.D.: Energy-efficient routing protocols in wireless sensor networks: a survey. IEEE Commun. Surv. Tutor. **15**(2), 551–591 (2013)
4. Abazeed, M., Faisal, N., Zubair, S., Ali, A.: Routing protocols for wireless multimedia sensor network: a survey. J. Sens. **2013**, 1–11 (2013)
5. Ehsan, S., Hamdaoui, B.: A survey on energy-efficient routing techniques with QoS assurances for wireless multimedia sensor networks. IEEE Commun. Surv. Tutor. **14**(2), 265–278 (2012)
6. Shu, L., Zhang, Y., Yang, L.T., Wang, Y., Hauswirth, M., Xiong, N.: TPGF: geographic routing in wireless multimedia sensor networks. Telecommun. Syst. **44**(1–2), 79–95 (2010)
7. Li, B.-Y., Chuang, P.-J.: Geographic energy-aware non-interfering multipath routing for multimedia transmission in wireless sensor networks. Inf. Sci. **249**, 24–37 (2013)
8. Spachos, P., Toumpakaris, D., Hatzinakos, D.: QoS and energy-aware dynamic routing in wireless multimedia sensor networks. In: 2015 IEEE International Conference on Communications (ICC), pp. 6935–6940. IEEE (2015)
9. Wang, K., Wang, L., Ma, C., Shu, L., Rodrigues, J.: Geographic routing in random duty-cycled wireless multimedia sensor networks. In: 2010 IEEE GLOBECOM Workshops, pp. 230–234. IEEE (2010)
10. Mohajerzadeh, A.H., Yaghmaee, M.H., Toroghi, N.N., Parvizy, S., Torshizi, A.H.: MREEP: a QoS based routing protocol for wireless multimedia sensor networks. In: 2011 19th Iranian Conference on Electrical Engineering, pp. 1–6. IEEE (2011)
11. Ukani, V., Kothari, A., Zaveri, T.: An energy efficient routing protocol for wireless multimedia sensor network. In: 2014 International Conference on Devices, Circuits and Communications (ICDCCom), pp. 1–6. IEEE (2014)
12. Cobo, L., Quintero, A., Pierre, S.: Ant-based routing for wireless multimedia sensor networks using multiple QoS metrics. Comput. Netw. **54**(17), 2991–3010 (2010)
13. Ghaari, A., Takanloo, V.A.: QoS-based routing protocol with load balancing for wireless multimedia sensor networks using genetic algorithm. World Appl. Sci. J. **15**, 1659–1666 (2011)
14. Almalkawi, I.T., Zapata, M.G., Al-Karaki, J.N.: A cross-layer-based clustered multipath routing with QoS-aware scheduling for wireless multimedia sensor networks. Int. J. Distrib. Sens. Netw. **2012**, 1–11 (2012)
15. do Rosário, D., Costa, R., Paraense, H., Machado, K., Cerqueira, E., Braun, T.: A smart multi-hop hierarchical routing protocol for efficient video communication over wireless multimedia sensor networks. In: 2012 IEEE International Conference on Communications (ICC), pp. 6530–6534. IEEE (2012)
16. Agarkhed, J., Biradar, G.S., Mytri, V.D.: Energy efficient QoS routing in multi-sink wireless multimedia sensor networks. Int. J. Comput. Sci. Netw. Secur. (IJCSNS) **12**(5), 25 (2012)
17. Jaiswal, A., Rao, S., Shama, K.: Application aware energy efficient geographic greedy forwarding in wireless multimedia sensor networks. Int. J. Eng. Adv. Technol. **1**(5), 222–227 (2012)

18. Zhang, L., Hauswirth, M., Shu, L., Zhou, Z., Reynolds, V., Han, G.: Multi-priority multi-path selection for video streaming in wireless multimedia sensor networks. In: Sandnes, F.E., Zhang, Y., Rong, C., Yang, L.T., Ma, J. (eds.) UIC 2008. LNCS, vol. 5061, pp. 439–452. Springer, Heidelberg (2008). doi:10.1007/978-3-540-69293-5_35

19. Bennis, I., Fouchal, H., Zytoune, O., Aboutajdine, D.: An evaluation of the TPGF protocol implementation over NS-2. In: 2014 IEEE International Conference on Communications (ICC), pp. 428–433. IEEE (2014)

20. Medjiah, S., Ahmed, T., Krief, F.: AGEM: adaptive greedy-compass energy-aware multipath routing protocol for WMSNS. In: 2010 7th IEEE Consumer Communications and Networking Conference, pp. 1–6. IEEE (2010)

21. Felemban, E., Lee, C.-G., Ekici, E.: MMSPEED: multipath multi-speed protocol for QoS guarantee of reliability and timeliness in wireless sensor networks. IEEE Trans. Mob. Comput. **5**(6), 738–754 (2006)

22. Levis, P., Gay, D.: TinyOS Programming. Cambridge University Press, Cambridge (2009)

Ensuring Connectivity in Wireless Sensor Network with a Robot-Assisted Sensor Relocation

Sahla Masmoudi Mnif$^{(\boxtimes)}$ and Leila Azouz Saidane

National School of Computer Science, University of Manouba, 2010 Manouba, Tunisia
sahla.masmoudi@ensi-uma.tn, leila.saidane@ensi.rnu.tn

Abstract. Wireless sensor networks (WSN) are used to survey a given Region Of Interest (ROI) especially the WSNs are used to survey hazardous and unreacheable zones like military zone or frontiers, the survey of this kind of areas is very important and can prevent from terorrist events. Ensuring connectivity between all deployed sensors of the (ROI) is a challenging issue. Random technique of node deployment such as stochastic node dropping result in hole creating in some areas of the network and redundant nodes may appear in other areas. In this paper, we propose two assisted-robot algorithms in which we use redundant sensors and relocate them in order to cover holes. We exploit here the redundancy of sensors to connect the formed partitions of sensors. We propose two strategies for the robot functioning and sensor relocation, the first strategy is a grid based one, in this solution the controlled area is divided into a virtual grid and the robot movement is based on this grid, we called this strategy "Grid-Based Walk with Memorization" (GBWM). The second strategy is an island based strategy, the network is composed of a set of disconnected island and the task of the robot is to connect the formed islands, we called this strategy "Island-Based Walk with Memorization", noted (IBWM). Through extensive simulations we show the importance of these algorithms.

Keywords: Wireless sensor network · Relocation · Connectivity · Mobile robot · Redundant nodes

1   Introduction

Wireless Sensor Network (WSN) consists of a set of small entities characterized by a limited amount of energy and low computational capabilities. WSN is designed basically to survey a given area in order to detect a specific phenomenon (intrusion, seism, fire...). In order to ensure this task, connectivity and coverage must be guaranteed in the network.

Generally, the initial deployment of sensors over the network is done in a random way, particularly in the hazardous areas. This kind of deployment leads in most of cases to a partitioned network. Hence, the total connectivity is not

© Springer International Publishing AG 2016
S. Boumerdassi et al. (Eds.): MSPN 2016, LNCS 10026, pp. 109–121, 2016.
DOI: 10.1007/978-3-319-50463-6_10

ensured and is causing in consequence a faulty network. Therefore, the sensors must be relocated to achieve the connectivity and to enhance the coverage in the WSN. Many solutions have been proposed in recent years to redeploy the sensors in order to ensure the connectivity and to improve the coverage over the network. One trivial solution is to provide to deployed sensors motion capabilities so that they will be able to relocate themselves and enhance the connectivity and the coverage in the network.

However, the mobility task requires an excessive consumption of energy of nodes which cause depletion of sensors and which leads to a lack of both connectivity and coverage.

One relevant solution for sensor relocation is to use the "partial" motion feature of nodes. In this context, the network is heterogeneous, containing standard fixed sensors and mobile entities that may have more important level of energy and higher processing capabilities.

These mobile entities are called Actuators or simply "Actors". In our work, we deal with a heterogeneous WSN consisting of a set of fixed sensors deployed in a random way in the area under control and a mobile robotused as actor in the network. The random deployment leads generally to partitioned networks with unguaranteed connectivity and coverage.

The robot is used to assist the network and to improve its performance. The main role of the robot is to exploit the redundant sensor nodes and relocate them to ensure the connectivity.

Our paper is structured as follows: We start with a brief review of the existing solutions for the redeployment of sensors in WSN. Then we present our proposed solutions and we validate them by means of different experiments and we close this paper by conclusion.

2   Related Work

In recent years, sensor relocation has been a challenging matter that was studied by many researchers. Several solutions have been proposed to solve the redeployment issue. One relevant solution was to provide motion capability to all sensors. This way, the sensors can move and relocate themselves in order to adjust the topology and achieve the connectivity and/or the coverage.

The sensors must synchronize their movement to enhance the network topology. Among the proposed solutions, we mention particularly the cascade motion which is detailed in [2]: instead of moving directly to the target, the sensor nodes adopt a cascade movement which means that the nearest node to the target point will move there, and the location of nearest node is replaced by moving another sensor and so on. Virtual Forces Aspect has been also proposed as a solution for sensor relocation. In this way, deployed sensors communicate together and compute their new locations in order to ensure connectivity and/or coverage. Then these sensors exerce a repulsive or an attractive force to move to their estimated locations. This strategy was studied and presented in [18].

The mobility of nodes is very efficient and improves the network topology, but it requires an important energy consumption which causes the node depletion and decreases the network lifetime.

Other solutions consist of the use of fixed sensor nodes and the network is assisted by some "actors" like mobile robots. Some studies, proposed to use the robot to carry data between disconnected sensors so that robot collects the detected event from nodes and then delivers these informations for the other nodes. This approach is presented by [19]. In this way the event is delayed and a latency time is introduced which can be considered as a shortcoming for critical applications. In other proposed solutions, the actors are advanced mobile sensors that exploit the redundant nodes and relocate them to achieve better connectivity and/or coverage trying to preserve the network lifetime as long as possible.

3   Proposed Solution

In our work we propose to survey hazardous and unaccesible zones like military fields, frontiers, unreacheable forests. For this kind of areas, the deterministic deployment is impossible. Hence we consider an initial random deployment of sensor nodes; we scattered a large amount of sensors within the region of interest. Each node in the network knows its own position by attached GPS or other technique of positioning. Sensors have the same communication range r_c and the same sensing range r_s, we note that $r_c \geq r_s$. Using this kind of sensor deployment, connectivity in the resulted network is not guaranteed and the network will be partitioned. Furthermore, some sensors will be redundant and some other sensors will be completely disconnected. Our idea is to relocate redundant sensors in order to improve the connectivity in the network. For this purpose and while our sensors are static, we propose to use a mobile robot in order to relocate redundant nodes.

3.1   Characteristics of Robot

To relocate redundant sensors we use a robot which is described as follows:

- the robot has a motion capacity and can move to any fixed direction with a static speed.
- the robot has a very important storage and computation capacities.
- the robot has a communication range noted R_c larger than the other static sensor nodes.
- the robot is able to locate its position.
- the robot knows the limits (frontiers) of the region of Interest.
- Initially, the mobile robot is equipped by sensor nodes (Reserve Nodes) that it can use them to cover holes.
- The robot knows the numbers of the deployed static nodes.
- The robot knows the position of the "Sink" node in the network.
- Initially, the robot has no idea about the network topology.

3.2 Robot Functioning

The main task of the robot is to improve the connectivity in the network. For this purpose the robot deals with redundant nodes and relocate them to enhance the connectivity in WSN.

To ensure their tasks, the robot can:

- move along the region of interest.
- pick up redundant nodes.
- communicate with static sensors.
- relocate static sensors.

We note that the robot had no idea about the network topology, so that it should move through all the area in order to discover the topology of the network (position of static sensors, position of redundant nodes, number of redundant nodes,...) and at the same time, it tries to improve the connectivity by relocating redundant nodes. Hence the robot should organize its motion and its functioning in order to achieve the connectivity in a WSN as soon as possible.

In our work we propose two strategies for the robot movement and sensor relocation. The first strategy is a grid-based one, the area will be partitioned in a virtual grid and based on this partitioning the robot will move and relocate redundant nodes we called this strategy "Grid-Based Walk with Memorization" (GBWM). The second strategy is an island-Based strategy in which the deployed network will be partioned in a set of disconnected islands and the main task of the robot is to connect islands using redundant nodes, we called this strategy "Island-Based Walk with Memorization" and noted by (IBWM).

In our work we will present this two strategies and then establish a comparison between them.

4 Grid-Based Walk with Memorization

In this strategy the controlled area is divided through a virtual grid, when we use a random deployment of sensors, some cells may be empty and so that there will be some disconnected cells. The main task of the robot is to connect the disconnected cells to the cell containing the "sink" node.

4.1 Network Partitioning

We propose to divide the controlled area according to rectangular (or a square) grid. The length of each side of a cell is fixed and is the same in all the network. The length of each side is less than he communication range r_c of a sensor. Hence sensors belonging to the same cell are able to communicate together so sensors in the same cell are connected. Figure 1 shows an example of a grid-based network partitioning.

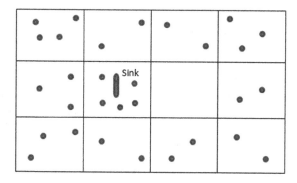

Fig. 1. Grid-based partitioning

4.2   Grid-Head Election

To facilitate the communication between nodes and robot, we propose to choose a grid head in each cell, the role of this grid head is to:

- communicate with all sensors of its cell.
- collect all information about its cell (redundant nodes, position of nodes...).
- communicate with the robot.

The Grid-head is elected as the node:

- having the highest level of residual energy.
- in case of multiple candidates, the node having more neighbors is elected.
- in case of multiple candidates, the node with the highest identity is elected.

Upon its election, the Grid-Head announces its presence in its cell. The Grid-Head collects all informations concerning its cell in order to forward them to the robot when needed.

4.3   Redundant Nodes Identification

A node is said to be redundant if its perception cell is fully covered by other cells of the other sensors. All information concerning redundant nodes are recorded in the Grid-Head and all redundant nodes go to the sleeping mode in order to economise energy.

When needed, the Grid-head can awaken some redundant nodes and update the list of the recorded redundant nodes.

4.4   Robot Functionning

The robot is able to move horizontally or vertically through the virtual grid. Initially, the robot moves according to the grid.

Periodically, the robot stops (after a fixed distance) and sends a Hello-Robot message indicating its position, its identity in order to discover the network topology. Each Grid Head receiving this message, has to reply to the robot with all information concerning its cell.

When the robot receives these informations from grid-heads, it detects the presence of sensors and then it discovers a part of the network topology. Hence the robot discovered a set of cells of the network.

If the robot encounters an empty cell or a disconnected cell, it tries to connect these cells to the cell containing the "sink" node. All discovered informations are memorized by the robot.

For the GBWM strategy, the robot:

- collects redundants nodes if it has the possibility to do.
- if it encounters an empty cell or a disconnected cell, the robot computes the shortest path (according with the grid) to achieve the sink node.
    - if the robot has the number of needed sensors, the robot connects this cell to the sink node.
    - if the robot has not the needed number of nodes it decides to collect the nearest redundant nodes (using the memorized informations).
- when a robot encounters a redundant node it picks up this sensor, if its reserve is not full, otherwise, the robot memorizes the position of the redundant nodes.

5   Island-Based Walk with Memorization

Using this kind of sensor deployment, connectivity in the resulted network is not guaranteed. Furthermore, the random deployment leads to the creation of disconnected islands. We call "Island" a set of connected sensors.

The Island containing the **"Sink"** is called **Main Island**. In each island, the connectivity is ensured but the islands are not able to communicate between each other. Generally, each island contains redundant sensor nodes.

5.1   Redundant Sensor Identification

We use a hexagonal partition of region in order to identify and locate redundant sensors. The sensors belonging to two adjacent cells are able to communicate. A sensor is said to be redundant if its cells (perception zone) is covered by other cells of the other network.

In order to economize energy end to reduce its consumption, redundant nodes after their identification should enter in a sleep mode. When a redundant node is deployed or is relocated, it should be awaken (in active mode).

5.2   Island-Head Identification

For each Island, a chief is elected, called Island-Head, this island-Head collects all the information about the island (positions of redundant nodes, positions of nodes in the islands...).

The Island-head is elected as the node with the highest level of residual energy and the largest set of neighbors. In order to select the Island-Head, an election factor denoted by f is defined by this equation:

$$f = \frac{1}{2} * \frac{E_{res}}{E_{max}} + \frac{1}{2} * \frac{Nb_n}{Nb_{nodes}}$$

where E_{res} and E_{max} represent respectively the residual energy and the maximum level of energy for a given node. Nb_n and Nb_{nodes} refers to the number of neighbors of a sensor node and the numbers of nodes in a given Island.

The node with the highest value of f is elected as an Island-Head. In case of multiple candidates, the node with the highest Identity is elected.

A backup Island-head is chosen to replace Island-Head in case of its depletion. The backup Island-Head is the second candidate after the Island-Head. When the Island-Head is elected, it collects the positions of redundant nodes and these nodes go to the passive mode (sleeping mode) to save energy of the whole network.

The robot Knows the position of the sink node, therefore, it has no idea on the network topology. Hence, the main role of the mobile robot is to discover the topology of the network and simultanously it tries to redeploy redundant sensors in order to enhance the network topology and to ensure connectivity between each Island of the network and the main Island to obtain a connected network.

We notice that, the mobile robot is considered as a sophisticated entity with an important computational capability and a large amount of energy. We suppose also that the robot can be recharged as needed. The robot has also sensing and communication capabilities, we noted by R_c the communication range of the robot and R_s its sensing range; $R_c \leq R_s$. We assume also that the robot is equipped by a number of sensors Nb_{res} that can be used to heal connectivity holes.

Each couple of nodes (whether sensor node or robot) can communicate directly when they are within each other communication range. In our solution, we will exploit sensor redundancy: the mobile robot picks up redundant nodes and relocates them to enhance connectivity. We propose a new strategy called Island-Based Walk with Memorization (IBWM) in which the robot walk is made based on the recently discovered informations.

Periodically, the robot stops (after a distance of $2 * R_c$ and sends a Hello-Robot Message. Each sensor receiving a "Hello-Robot", forwards this message to its "Island-Head". The "Island-Head" replies with "Island-information" message containing all the informations concerning this island (position of nodes, positions of redundant sensors, sensors identities, number of redundant nodes...).

- if the robot does not receive any reply, it continues its walk in a random direction.
- when a robot encounters an Island, it memorizes all the informations concerning this Island mainly the locations of redundant nodes.
- if the robot receives an "Island-information" message, it computes the position of the nearest node of the "sink" and then it calculates the number of needed sensors to connect the island to the main Island.

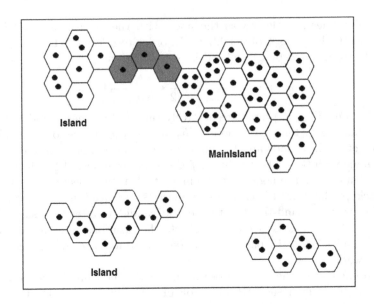

Fig. 2. Connecting two Islands (Color figure online)

- if this requested number of sensors is available on the robot, it relocates them (the nodes will be relocated according to hexagonal pavement).
- when they are no carried sensors on the robot, this robot returns back to the nearest redundant nodes and picksup them after that, it relocates them like in the IBERIA algorithm. Then the continues its travel in a random manner.

Figure 2 describes the way to connect an Island to the Main-Island. Green nodes are placed and redeployed by the mobile robot.

6 Performance Evaluation

Our proposed solution is implemented under NS2 simulator. Several simulations were established with different scenarios. For all simulations we use a large number of deployed sensors is quite large to ensure full connectivity and coverage over the network.

Sensors are initially deployed randomly through a square ROI, we set $r_c = 40\,m$, $r_s = 25\,m$ and $R_c = 60\,m$. We set the dimension of the ROI to $400 * 400\,m^2$. The number of deployed sensor nodes is in a first time fixed to 100 sensors. In a second step, we will vary the number of deployed sensors from 350 to 750 sensors.

The number of formed islands over the network is a primordial factor which gives us an idea on the total connectivity in the network. Figure 3 shows that the number of formed islands increases when the number of deployed sensors

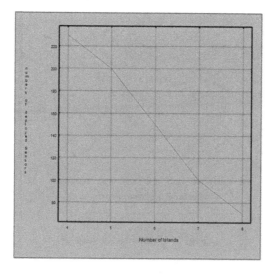

Fig. 3. Number of formed Islands

decreases. Connectivity between nodes increases when the number of deployed sensors increases.

The most important factor is the redundancy of sensors in the network.

Figure 4 shows that redundancy of nodes decreases when the number of Islands in the network increases.

When the number of Islands increases, the number of distributed sensors decreases and so the number of redundant nodes.

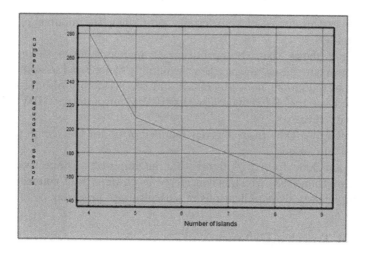

Fig. 4. Redundancy

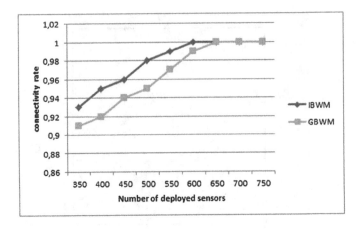

Fig. 5. Connectivity rate

After implementing the robot algorithm, through simulation we will evaluate some metrics such as total distance traveled by robot and connectivity rate.

6.1 Connectivity Rate

The connectivity rate (CR) is he average of connected sensors in the network, this metric should be maximized to enhance the performance of the tested algorithm. CR is giving by the following equation:

$$\frac{Number\ of\ connected\ islands\ to\ the\ Main\ Island}{Initial\ Number\ of\ Islands} \qquad (1)$$

Figure 5 shows the curves of connectivity rate when the number of deployed sensors increases.

This Fig. 5 shows that connectivity rate increases when the number of deployed sensors increases, in fact when the number of deployed sensors increases, the number of redundant nodes increases and the possibility of improving connectivity increases too. We remark also that IBWM outperforms GBWM in term of connectivity rate.

We remark also that from a given number of nodes the connectivity rate achieve a certain stability.

We can conclude from this figure that an important connectivity rate is ensured (more than 0.98) which is considered as an important connectivity rate for some application like applications of agriculture of precision.

6.2 Robot Traveled Distance

The traveled Distance is the distance traveled by a mobile robot after performing the algorithm.

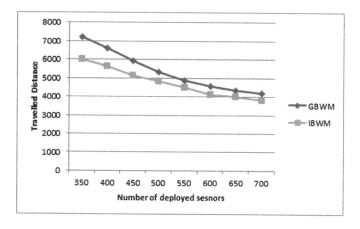

Fig. 6. Travelled distance by robot

Figure 6 shows the travelled distance by a mobile robot according to the number of sensors. We remark that the travelled distance decreases when the number of deployed sensors increases.

This can be explained by the fact that the number of holes decreases when the number of deployed sensors increases. The travelled distance achieve a certain stability when the number of deployed sensors increases and achieve a certain limit meaning that we can predict the minimum number of deployed senors.

7   Conclusion and Future Work

In this paper we proposed a robot-based sensor relocation to ensure connectivity in the Wireless sensors networks. We proposed two strategies to model our network. The first startegy is a Grid-based strategy and called Grid-Based Walk with Memorization GBWM which consists in partitioning the network in a virtual grid.

The second strategy is called Island-Based Walk with Memorization, IBWM, for this strategy we model our networks with a set of disconnected Islands and propose to connect them using a mobile robot. We proposed to use a mobile robot to relocate redundant nodes in order to improve the connectivity in the networks.

Through several simulations we validate our work. We conclude that Island based solution outperforms Grid-based One, especially in term of travelled distance and hence in terms of time. But from our work we can conclude that we can use Grid-based solution for a given application which do not have hard constraints like precision agriculture. Grid-based solutions are more suitable for this kind of applications especially that the partitioning of area in a grid is already made by the plants and the trees. Island Based solution can be used for more sophisticated solutions like detection intrusion in military areas or in a battle field, where the connectivity between all deployed nodes must be guaranteed.

Our work can be used also make a trade off between the number of deployed sensor and the requested performance of the network (the needed connectivity rate and or the travelled distance of the mobile robot).

As a further work we proposed to enhance these solutions to multi robot and to ensure a synchronization between robots.

References

1. Kershner, R.: The number of circles covering a set. Am. J. Math. **61**, 665–671 (1939)
2. La Porta, T., Wang, G., Cao, G., Wang, W.: Sensor relocation in mobile sensor networks. In: Infocom 2005 (2005)
3. Wang, F., Thai, M., Du, D.: On the construction of 2-connected virtual backbone in wireless networks. IEEE Trans. Wireless Commun. **8**, 1230–1237 (2009)
4. Abbasi, A., Younis, M., Akkaya, K.: Movement-assisted connectivity restoration in wireless sensor and actuator networks. IEEE Trans. Parallel Distrib. Syst. **20**, 1366–1379 (2009)
5. Wang, G., Cao, G., Laporta, T.: A bidding protocol for deploying mobile sensors. In: The 11th IEEE International Conference on Network Protocols (ICNP) (2003)
6. Wang, G., Cao, G., Laporta, T.: Movement-assisted sensor deployment. In: Infocom 2004 (2004)
7. Fletcher, G., Li, X., Nayak, A., Stojmenovic, I.: Carrier-based sensor deployment by a robot team. In: IEEE SECON (2010)
8. Mei, Y., Xian, C., Das, S., Hu, Y.C., Lu, Y.H.: Sensor replacement using mobile robots. Comput. Commun. **30**(13), 2615–2626 (2007)
9. Xuan, D., Yun, Z., Bai, X., Kumar, S., Lai, T.H.: Deploying wireless sensors to achieve both coverage and connectivity. In: Mobile Ad Hoc Networking and Computing (2006)
10. Li, X., Santoro, N.: ZONER: a zone-based sensor relocation protocol for mobile sensor networks. In: IEEE WLN (2006)
11. Li, X., Santoro, N., Stojmenovic, I.: Mesh-based sensor relocation for coverage maintenance in mobile sensor networks. In: Indulska, J., Ma, J., Yang, L.T., Ungerer, T., Cao, J. (eds.) UIC 2007. LNCS, vol. 4611, pp. 696–708. Springer, Heidelberg (2007). doi:10.1007/978-3-540-73549-6_68
12. Howard, A., Mataric, M.J., Sukhatme, G.S.: Mobile sensor networks deployment using potential fields: a distributed, scalable solution to the area coverage problem. In: Asama, H., Arai, T., Fukuda, T., Hasegawa, T. (eds.) Distributed Autonomous Robotics Systems, pp. 299–308. Springer, Heidelberg (2002)
13. Egea-Lpez, E., Vales-Alonso, J., Martnez-Sala, A.S., Pavon-Mario, P., Garca-Haro, J.: Simulation tools for wireless sensor networks. In: Summer Simulation Multiconference - SPECTS (2005)
14. Akyildiz, I.F., Su, W., Sankarasubramaniam, Y., Cayirci, E.: A survey on sensor networks. IEEE Commun. Mag. **40**, 102–114 (2002)
15. Watteyne, T.: Using existing network simulators for power-aware self-organizing wireless sensor network protocols. In: INRIA (2006)
16. Chalhoub, G.: Reseaux de capteurs sans fil. Clermont Universite (2009)
17. Gallais, A., Carle, J., Simplot-Ryl, D.: La k-couverture de surface dans les reseaux de capteurs. In: AlgoTel (2007)

18. Wang, X., Wang, S.H., Bi, D.: Virtual force-directed particle swarm optimization for dynamic deployment in wireless sensor networks. In: Huang, D.-S., Heutte, L., Loog, M. (eds.) ICIC 2007. LNCS, vol. 4681, pp. 292–303. Springer, Heidelberg (2007). doi:10.1007/978-3-540-74171-8_29
19. Zhao, W., Ammar, M., Zegura, E.: A message ferrying approach for data delivery in sparse mobile adhoc networks. In: Mobihoc 2004 (2004)

A New Non-intrusive Assessment Method for VoLTE Quality Based on Extended E-Model

Duy-Huy Nguyen$^{(\boxtimes)}$, Hang Nguyen, and Éric Renault

SAMOVAR, Télécom SudParis, CNRS, Université Paris-Saclay,
9 rue Charles Fourier, 91011 Evry Cedex, France
{duy_huy.nguyen,hang.nguyen,eric.renault}@telecom-sudparis.eu

Abstract. Voice over Long Term Evolution (VoLTE) is always a main service that brings a big benefit for mobile operators. However, the deployment of VoLTE is very complex, specially for guaranteeing of Quality of Service (QoS) to meet quality of experience of mobile users. The key purpose of this paper is to present an object non-intrusive prediction model for VoLTE quality based on LTE-Sim framework and the extended E-model. This combination allows overcoming the lack of the E-model is that how to determine exactly its inputs. Besides, we propose to complement the jitter buffer as an essential input factor to the E-model. The simulation results show that the proposed model can predict voice quality in LTE network via both the E-model and the extended E-model. The proposed model does not refer to the original signal, thus, it is very suitable for predicting voice quality in LTE network for many different scenarios which are configured in the LTE-Sim framework. The simulation results also show that the effect of jitter buffer on user perception is very significant.

Keywords: VoLTE · Voice quality · LTE · Extended E-model · MOS

1 Introduction

LTE network is developed by the Third Generation Partnership Project (3GPP) [8]. It is a mobile network which has high data rates, low latency and fully packet-based. It means to improve the capability of legacy system by increasing data rates and extending superior QoS for various multimedia applications. Voice over LTE network (called VoLTE) is a main service of LTE network. Since LTE network is a full packet-switched, thus, the deployment of VoLTE service is very complicated. All voice traffics over LTE network are VoIP (Voice over Internet Protocol) including VoLTE. According to [10], there are two types of voice traffic over LTE network, those are VoLTE and VoIP. VoLTE is really the VoIP service with QoS guaranteed [10]. Table 1 represents the service classes in LTE network where voice service is a Guaranteed Bit Rate (GRB) service which has the second priority just after IP Multimedia Subsystem (IMS) signaling. However, in order to guarantee VoLTE quality is an extreme challenge.

© Springer International Publishing AG 2016
S. Boumerdassi et al. (Eds.): MSPN 2016, LNCS 10026, pp. 122–136, 2016.
DOI: 10.1007/978-3-319-50463-6_11

Table 1. LTE service classes with QoS requirements

Resource type	Priority	Packet delay budget (ms)	Packet error loss rate	Example services
Guaranteed Bit Rate (GBR)	2	100	10^{-2}	Conversational voice
	4	150	10^{-3}	Conversational video (live streaming)
	3	50	10^{-3}	Real-time gaming
	5	300	10^{-6}	Non-conversational video (buffered stream)
Non-GBR	1	100	10^{-3}	IMS signaling
	6	300	10^{-6}	Video (buffered streaming) TCP-based (e.g. www, e-mail, chat, FTP, P2P sharing, progressive video, etc.)
	7	100	10^{-6}	Voice, Video (live streaming, Interactive Gaming)
	8	300	10^{-3}	Video (buffered streaming) TCP-based (e.g. www, e-mail, chat, FTP, P2P sharing, progressive video, etc.)
	9		10^{-6}	

In communications systems, the perceived voice quality is usually represented as the MOS. MOS can be attained by many methods. These methods are divided into two groups called subjective methods and objective ones. Subjective methods humans listening to a live stream or a recorded file and rating it on a ratio of 1 (poor) to 5 (excellent) [11]. These methods have some disadvantages such as too expensive, time consuming and are not suitable for a large network infrastructure. Otherwise, objective methods have more advantages, they eliminate the limitations of subjective methods. Objective methods are classified into two approaches: intrusive and non-intrusive ones. The intrusive methods (e.g. Perceptual evaluation of speech quality (PESQ) [13]) are more exact and are widely utilized to predict aware voice quality. However, they are not suitable for real-time services such as VoIP because they require original signals to refer. The non-intrusive methods (e.g. ITU-T E-model [12]) are computational models that are used for transmission planning purposes. They are not as accurate as the intrusive approaches and they do not have complex mathematical operations. The obtained results from objective methods do not always well relate to human perception. The main advantage of the non-intrusive methods are they predict voice quality without any reference to the original signals and they require less parameters than the intrusive methods. For several typical non-intrusive methods, authors in [20] proposed to use Random Neural Network (RNN) to assess voice quality over internet. Voice quality assessment was predicted in [14] using

RNN. Another non-intrusive method was proposed in [9] based on RNN for evaluating video quality in LTE network. Authors in [7] proposed to use Wideband (WB) E-model to predict VoLTE quality for minimizing redundant bits generated by channel coding. In [6], authors investigated the effects of PLR and delay jitter on VoIP quality to assess prediction errors of MOS for the E-model. A new model called "Packet-E-Model" was proposed in [16] to measure speech quality perception for VoIP in Wimax network. In [4], authors presented a voice quality measurement tool based on the E-model. A framework of objective assessment method for estimating conversational quality in VoIP was proposed in [22]. In [3], a simplified versions of the E-model were proposed to simplify the calculations and focus on the most important factors required for monitoring the call quality. According to our knowledge, at present, there are not any proposals which allow to predict VoLTE quality using LTE-Sim framework and extended E-model.

We see that, the lack of the E-model is how to measure exactly the its input parameters. Therefore, we need a method to measure the input factors of the E-model. In this paper, we propose to use the LTE-Sim [19] to calculate delay and PLR for voice users. LTE-Sim is a famous framework which allows to simulate entire LTE network, thus, it is quite similar to a real system. Outputs of LTE-Sim are Delay and PLR which are essential input parameters of the E-model. In addition, we also take effects of network jitter into account by considering jitter parameter (I_j) as an input parameter of the E-model. In order to obtain more real results, we simulate voice service with mobility in LTE heterogeneous network.

The rest of this paper is organized as follows: Overview of the system model is described in Sect. 2. In Sect. 3, we present the proposed model. The simulation results and performance evaluation of the proposed model are analysed in Sect. 4. The conclusion and future work is represented in Sect. 5.

2   The System Model

2.1   The LTE-Sim Framework

LTE-Sim is an open-source framework which is developed by Giuseppe Piro and his colleagues [19]. It is freely available for scientific community. It is used to simulate entire LTE network. There are many researchers who used LTE-Sim to simulate their proposals such as scheduling strategies, radio resource optimization, frequency reuse techniques, the adaptive modulation and coding module, user mobility, and etc. for both downlink and uplink directions and in both multicell/multiuser environments. The implemented protocol stack of LTE-Sim is represented on Fig. 1 for both user-plane and control-plane, thus, it is nearly similar to the real LTE system.

Figure 1 can be briefly described as follows: When a voice traffic flow transmitted over the LTE-Sim, it is encapsulated sequentially with network protocols. For the downlink direction, the VoLTE packet uses transport protocols of Real-time Transport Protocol (RTP), User Datagram Protocol (UDP) and Internet Protocol (IP). It is then packetized with radio protocols such as Packet Data

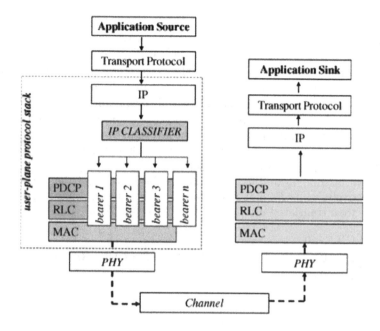

Fig. 1. The implemented protocol stack in LTE-Sim [19]

Convergence Protocol (PDCP), Radio Link Control (RLC) and Medium Access Control (MAC), and Physical (PHY) layer before it is transmitted over the air interface. LTE-Sim supports both IPv4 and IPv6 protocols with header sizes are 40 and 60, respectively while the voice payload is about 32 bytes, thus, to reduce the overhead, Robust Header Compression (RoHC) is deployed at PDCP layer. The IP header is then compressed by RoHC down to only 1–4 bytes, normally 3 bytes.

LTE-Sim allows to simulate a heterogeneous traffic over LTE network. Therefore, it is quite similar to a real system. The outputs of this software including many parameters where there are delay and PLR. LTE-Sim supports multi-cell/multiuser with mobility in a heterogeneous network. The details of the simulation scenario is represented in Sect. 3.

2.2 The Extended E-Model

E-model is a computational model developed and standardized by ITU-T [12]. It is used to estimate the MOS for narrow band audio quality. The output of the model is R-factor. The values of this R-factor in range of 0–100 where 100 is the best and 0 is the worst quality. And then, it is mapped to the corresponding MOS value. The standard R-factor in the E-model is defined as follows:

$$R = R_0 - I_s - I_d - I_{ef} + A \tag{1}$$

In which: R_0: The basic signal-to-noise ratio which consists of noise sources such as circuit and room noise. In this model, its value is set to 94.2. I_s: The simultaneous impairment factor, it is the sum of all impairments which may occur more or less simultaneously with the voice transmission. In this model, the default value is set to 0. I_d: The delay impairment factor, representing all impairments due to delay of voice signals. I_{ef}: The equipment impairment factor, capturing the effect of signal distortion due to low bit rates of the codec and packet losses of random distribution. A: The advantage factor, capturing the fact that some users can accept a reduction of quality due to the mobility of cellular networks. In this model, this factor is set to 0.

In above factors, I_d and I_{ef} are affected by end-to-end delay and packet loss, respectively, while R_0 and I_s do not depend on network performance. The R-factor is then translated into the MOS as follows [12]:

$$MOS = \begin{cases} 1, \text{if } R < 0 \\ 1 + 0.035 \times R + 7 \times 10^{-6} \times R \times (R - 60) \times (100 - R), \text{if } 0 \le R \le 100 \\ 4.5, \textbf{otherwise} \end{cases}$$

(2)

The relation between R-factor, user perception, and MOS is described in the Table 2.

Table 2. R-factor and MOS with corresponding user satisfaction

R	User satisfaction	MOS
$90 \le R < 100$	Very satisfied	4.3–5.0
$80 \le R < 90$	Satisfied	4.0–4.3
$70 \le R < 80$	Some users dissatisfied	3.6–4.0
$60 \le R < 70$	Many users dissatisfied	3.1–3.6
$50 \le R < 60$	Nearly all users dissatisfied	2.6–3.1
$R < 50$	Not recommended	<2.6

After setting the default values for the E-model, Eq. (1) can be rewritten as follows:

$$R = 94.2 - I_d - I_{ef}$$

(3)

We see that, when voice packet transmitted over an IP network, it is affected by many network impairments such as PLR, delay, jitter, etc. In the E-model, there is no presence of network jitter. In order to improve use satisfaction, we propose to add the I_j factor to the E-model. It's obvious, when add I_j to the E-model, user perception of the extended E-model will be lower than E-model, and the E-model can be described as the following formula:

$$R = 94.2 - I_d - I_{ef} - I_j$$

(4)

Equation (4) shows that the R-factor depends on end-to-end delay (I_d), total loss probability (I_{ef}), and network jitter (I_j). Hence, in order to compute the R-factor, we must to count these factors. The I_d is a factor which is affected by end-to-end delay and is calculated as follows [18]:

$$I_d = 0.024 \times d + 0.11 \times (d - 177.3) \times H(d - 177.3) \tag{5}$$

In which: $H(x)$ is the Heavyside function:

$$H(x) = \begin{cases} 0, \text{if } x < 0 \\ 1, \textbf{otherwise} \end{cases} \tag{6}$$

In Eq. (5), d represents the total end-to-end delay (or mouth-to-ear delay) of speech packet. It is one of output results of LTE-Sim software. The I_{ef} is determined according to packet loss. In order to compute this factor, we use the equation in [17] as follows:

$$I_{ef} = \lambda_1 + \lambda_2 \times ln(1 + \lambda_3 \times e_l) \tag{7}$$

Where: The λ_1 represents the voice quality impairment factor caused by the encoder, λ_2 and λ_3 represent the effect of loss on voice quality for a given codec. Such that, these factors depend on the voice codec used. In this study, we use LTE-Sim [19] to simulate. This simulation tool supports only G.729 codec, thus, for this codec, the factors above has values as follows: $\lambda_1 = 11, \lambda_1 = 40, \lambda_3 = 10$. While e_l is the total loss probability (consisting of network and buffer layout) which has the value in range of 0..1. This factor is also obtained from the output results of LTE-Sim software.

The I_j represents the impacts of network jitter to voice quality. It also depends on the voice codec. In this paper, we use the method proposed in [6] as follows:

$$I_j = C_1 \times H^2 + C_2 \times H + C_3 + C4 \times exp(\tfrac{-T}{K}) \tag{8}$$

In which: C_1, C_2, C_3, C_4 are coefficients, K is time instant. These factors depend on the voice codec, for the G.729 codec, these factors have the values as follows: $C_1 = -15.5, C_2 = 33.5, C_3 = 4.4, C_4 = 13.6, K = 30$. The factor of T is the fixed buffer size of the voice codec. For the G.729 codec, the packet size is 20 ms, thus, normally $T = \propto \times 20$ where $\propto = 2, 3, 4, 5, 6, etc.$ The H is a factor of Pareto distribution and in range of 0.55 to 0.9. According to [6], the MOS slightly drops when H increases and it does not affect significantly on MOS score, thus, in this study, we select H = 0.6 for the simulation.

The final expression of the R-factor when utilizing the G.729 codec is described in Eq. (9).

$$\begin{aligned} R = 64.28 - [0.024 \times d + 0.11 \times (d - 177.3) \times H(d - 177.3)] - \\ 40 \times ln(1 + 10 \times e_l) - 13.6 \times exp(\tfrac{-4}{3}) \end{aligned} \tag{9}$$

The R-factor is then mapped to the MOS via Eq. (2). MOS reflects the perception of user. It is obvious, the higher MOS, the higher user satisfaction and inverse. How to determine exactly the perception of voice user is a big challenge of mobile operators.

3   The Proposed Model

Voice over LTE network (VoLTE) is a real-time service, and is fully deployed over an IP network, thus, in order to ensure VoLTE quality is very complex. There are very little methods that allow to monitor and to predict VoLTE quality and most of them are subjective ones which have to refer to the original signal, thus, they are not suitable for real-time services such as VoIP, Video, Gaming, etc. In this paper, we propose a new non-intrusive voice quality assessment method which based on LTE-Sim software [19] and the extension of the E-model [12]. LTE-Sim is a software which allow to simulate VoIP flow that is quite similar to real flow. Therefore, we see that, the simulation results are quite exact. The E-model is a computational model which allows to predict voice quality when it is transmitted from source to destination. However, in this model, there is not presence of network jitter. For VoIP flow, network jitter affects significantly on voice quality, thus, we propose to add the effects of network jitter to E-model via I_j factor. When the I_j is added to the E-model, the value of R factor will be lower, it leads to the lower user perception when compared to the standard E-model. The proposed model is represented on Fig. 2.

The principle of the proposed model as follows: Input parameters of the simulation scenarios are firstly fed to the LTE-Sim software. After finishing the simulation, we receive simulation results. In these results, we select Delay, PLR according to the number of user. These factors combining jitter buffer (JB) are the input parameters of the extended E-model. The output of E-model is R factor, and then it is mapped to MOS score. This MOS score performs the user satisfaction. The steps of the proposed model can be described as follows:

- **Step 1:** Setting input parameters of the simulation scenario.
- **Step 2:** Simulating the scenario in the LTE-Sim software.
- **Step 3:** Collecting the delay, PLR results from Step 2, and selecting values of JB.
- **Step 4:** Feeding the input parameters taken from Step 3 to the Extended E-model which is programmed in Matlab software [15] as follows:
  - Calculating R factor of the extended E-model according to Eq. (9).
  - R is then mapped to MOS score based on Formula (2).

After Step 4, we will receive the user satisfactions according to the number of user.

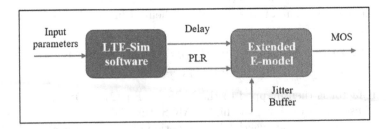

Fig. 2. The proposed implement model

LTE is a packet-switched network, thus, in order to simulate it as a real system, we select input data flows including a VoIP, a Video and a non real-time flow. In addition, the mobility is also included in our scenario, specifically, we select the speed of user is 30 and 120 km/h. We select the M-LWDF scheduler which is very suitable for real-time services. Besides predicting the user perception, we assess the effects of the Delay, PLR on user satisfaction according to the number of user.

4 Simulation Environment and Performance Evaluation

4.1 Simulation Environment

Traffic Model: In our scenario, the eNB is located at the center of the macrocell using an ommi-directional antenna in a 10 MHz bandwidth. Each UE uses a VoIP flow, a Video flow, and a INF-BUF flow at the same time. For the VoIP flow, a G.729 voice stream with a bit-rate of 8 kbps was considered. The voice flow is a bursty application that is modelled with an ON/OFF Markov chain [5]. For the video flow, a trace-based application that generates packets based on realistic video trace files with a bit-rate of 242 kbps was used in [21] and it is also available in [19]. In order to obtain a realistic simulation of an H.264 SVC video streaming, we used an encoded video sequence "*foreman.yuv*", which is publicly available.

The LTE propagation loss model is formed by four different models including: Path loss, Multipath, Penetration and Shadowing [1].

- Path loss: $PL = 128.1 + 37.6 \times log(d)$, with d is the distance between the UE and the eNB in km.
- Multipath: Jakes model
- Penetration loss: 10 dB
- Shadowing: Log-normal distribution with mean 0 dB and standard deviation of 8 dB.

Simulation Parameters: In this paper, we use M-LWDF scheduler for simulating. This scheduler is very suitable for voice traffic flow. The simulation process is performed in a single cell with interference with the number of users in the interval [10, 50] which move randomly at a speed of 30 and 120 km/h. The other basic parameters used in the simulation are represented in the Table 3.

4.2 Performance Evaluation

In order to simulate the traffic model in LTE-Sim, we use Modified Largest Weighted Delay First (M-LWDF) [2] scheduler with mobility in LTE heterogeneous network. The M-LWDF scheduler is very suitable for real-time services such as VoIP, Video, Gaming, etc. The analyses of the simulation results are represented in the following subsections.

Table 3. Simulation parameters

Simulation parameters	Values
Simulation duration	100 s
Frame structure	FDD
Cell radius	1 km
Bandwidth	10 MHz
Video bit-rate	242 kbps
VoIP bit-rate	8 kbps
User speed	30, 120 km/h
Number of user	10, 20, 30, 40, 50 UEs
Maximum delay	0.1 s
Packet scheduler	M-LWDF
Traffic model	VoIP, Video, and INF-BUF

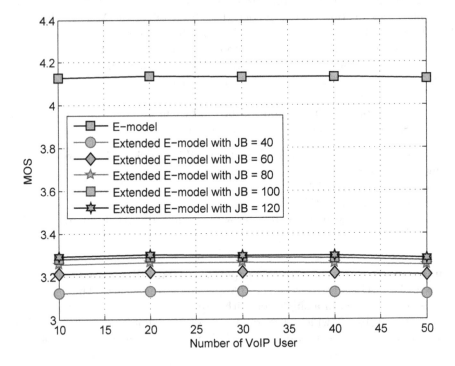

Fig. 3. MOS vs number of VoIP user at the speed of 30 km/h

Effects of Jitter Buffer on Voice Quality: In order to evaluate the effects of JB on voice quality, we set values of this factor of 40, 60, 80, 100, and 120. Figure 3 shows the relationship of the number of user (NU) vs user perception (MOS score) when the speed is 30 km/h. It's clear that, in this case, the

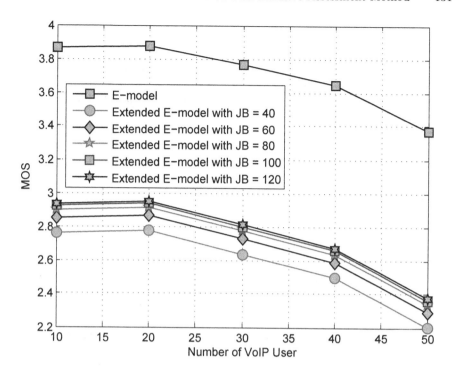

Fig. 4. MOS vs number of VoIP user at the speed of 120 km/h

E-model has the highest MOS score while the proposed model has the lower MOS. This is obvious because our proposed model considers effect of network jitter on the E-model. All cases have the slightly degraded MOS when NU increases. This figure also shows that the higher JB, the higher MOS. However, in fact, JB should not be too high. As shown on this figure, when JB is more than 80, the MOS increase not significantly. This is quite clear when JB corresponds to 100 and 120. It can be said that, when JB is more than 100, its effect on user perception is not significant.

Figure 4 shows the effects of JB when the speed is 120 km/h. It's clear that, when the speed is high, the user satisfaction is decreased clearly. All cases have the MOS score heavily decreasing when NU increasing. It's obvious, but the reduction is quite different from the case of the speed of 30 km/h. It's similar to the previous case, the higher JB, the higher user perception. The JB is directly proportional to user satisfaction while the speed is inverse.

Effects of Delay on Voice Quality: Figure 5 shows the relationship between delay vs user perception according to NU. It's clear that, the delay increases when NU increase. This leads to lower MOS score. The E-model has the highest MOS while the Extended E-model has the lower one. It can be said that, the higher NU and the higher delay, the lower MOS score (or lower user perception).

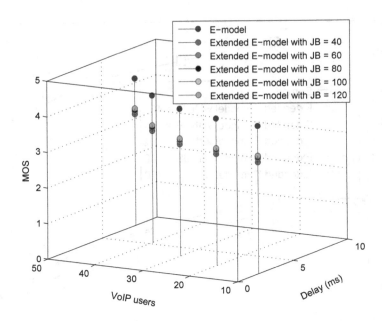

Fig. 5. Effects of delay on voice quality at the speed of 30 km/h

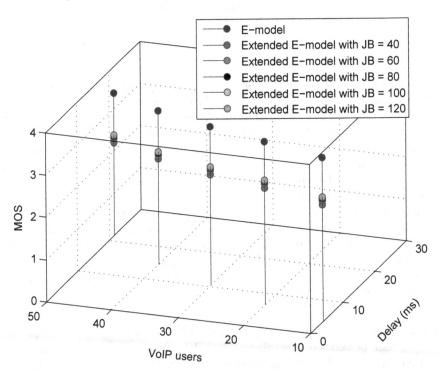

Fig. 6. Effects of delay on voice quality at the speed of 120 km/h

This principle is suitable for both the E-model and the Extended E-model (for all cases of JB). The detailed results of this case are represented in Table 4.

It's similar to the case of delay for the user speed of 30 km/h, the effects of delay on user satisfactions when the user speed is 120 km/h is shown on Fig. 6. The principle is similar to the user speed of 30 km/h, however, the MOS score is lower and heavily decreases when the NU or delay increases. This is right in both E-model and Extended E-model. The detailed results are described in Table 4.

Effects of PLR on Voice Quality: Figure 7 illustrates the effects of delay on user perception according to NU. It's clear that, MOS increases when the NU or PLR increases. This is suitable for both E-model and extended E-model. Normally, PLR increases when NU increases, but there is special case for the NU equals 10. This may be due to the system is usually not stable in fact. For both E-model and Extended E-model, the higher NU and the higher PLR, the lower MOS score. The E-model has the higher MOS when compared to the Extended E-model. For the Extended E-model, rule for the higher JB, the higher MOS is still met. For the detailed results, refer to Table 4.

For the case of the user speed equals 120 km/h as shown in Fig. 8, the principles for the case of user speed of 30 km/h are still met. The difference here is that in this case, the user perception decreases in comparison with the user

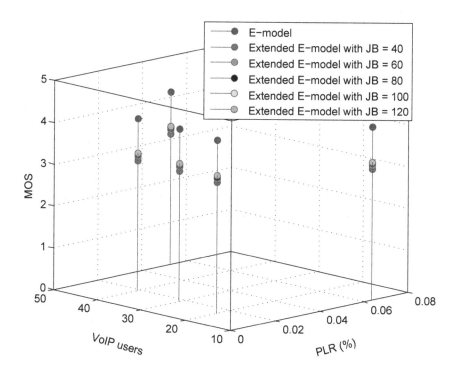

Fig. 7. Effects of PLR on voice quality at the speed of 30 km/h

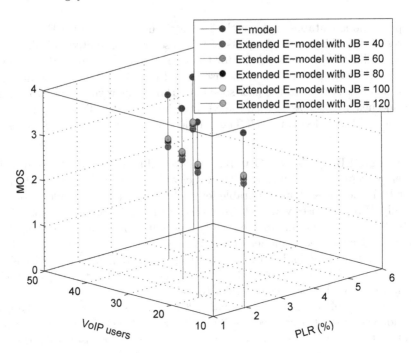

Fig. 8. Effects of PLR on voice quality at the speed of 120 km/h

Table 4. The detailed results of Figs. of 5–8

NU	Delay (ms)	PLR (%)	MOS					
			E-model	Extended E-model				
				JB = 40	JB = 60	JB = 80	JB = 100	JB = 120
User speed: 30 km/h								
10	1.94	0.064	4.1287	3.1204	3.21	3.2556	3.2789	3.2909
20	2.47	0.014	4.1351	3.13	3.2195	3.2651	3.2884	3.3003
30	3.45	0.017	4.1339	3.1282	3.2177	3.2633	3.2886	3.2985
40	5.26	0.018	4.1322	3.1257	3.2153	2.2609	3.2842	3.2961
50	8.03	0.053	4.1252	3.1151	3.2048	3.2504	3.2738	3.2857
User speed: 120 km/h								
10	4.04	1.943	3.8643	2.7591	2.851	2.8981	2.9223	2.9347
20	6.64	1.833	3.8772	2.7753	2.8672	2.9143	2.9385	2.9508
30	10.36	2.618	3.7632	2.6352	2.7271	2.7743	2.7986	2.811
40	14.74	3.448	3.6434	2.496	2.5874	2.6345	2.6587	2.6711
50	21.54	5.432	3.371	2.2044	2.293	2.3389	2.3626	2.3748

speed of 30 km/h and it heavily decreases when the NU or PLR increases. This is fully suitable. The detailed results of this case are represented in Table 4.

5   Conclusion

In this paper, we propose a new non-intrusive assessment method of voice quality over LTE network. The proposed model is the combination of the LTE-Sim framework and the extension of the E-model. The simulation results show that, the proposed model has the lower user perception in comparison with the E-model one. This is because in the proposed model, we take into account effects of network jitter. This means the effect of JB on user perception is very significant. The simulation results also show that user perception decreases when the user speed increases. For the case of user speed of 30 km/h, the user satisfaction slightly reduces when NU raises while it heavily decreases for the user speed of 120 km/h. For both cases of the user speed, the user perception increases when the JB increases. However, when JB is more than 100, the user satisfaction increases not significantly. For the delay and PLR, both increase when NU increases and meet the principle: the higher delay and PLR, the lower user satisfaction. It can be concluded that, the proposed model is very suitable for predicting VoLTE quality be cause it does not refer to the original signal. The user perception can be not as exact as the intrusive methods such as PESQ but it is quite simple, calculate very quickly. It is also very suitable for purpose of transmission planning and for voice assessment in laboratory for academic community and researchers. The proposed model can be used to predict VoLTE quality for many different scenarios which can be configured in the LTE-Sim. In the future, we will evaluate the proposed model for more NU, other voice codecs, and complement more parameters into the E-model.

References

1. 3GPP: Requirements for Evolved UTRA (E-UTRA) and Evolved UTRAN (E-UTRAN). TR 25.913, 3rd Generation Partnership Project (3GPP), December 2009. http://www.3gpp.org/ftp/Specs/html-info/25913.htm
2. Ameigeiras, P., Wigard, J., Mogensen, P.: Performance of the M-LWDF scheduling algorithm for streaming services in HSDPA. In: 2004 IEEE 60th Vehicular Technology Conference, VTC2004-Fall, vol. 2, pp. 999–1003. IEEE (2004)
3. Assem, H., Malone, D., Dunne, J., O'Sullivan, P.: Monitoring VoIP call quality using improved simplified E-model. In: 2013 International Conference on Computing, Networking and Communications (ICNC), pp. 927–931. IEEE (2013)
4. Carvalho, L., Mota, E., Aguiar, R., Lima, A.F., Barreto, A., et al.: An e-model implementation for speech quality evaluation in VoIP systems. In: Null, pp. 933–938. IEEE (2005)
5. Chuah, C.N., Katz, R.H.: Characterizing packet audio streams from internet multimedia applications. In: IEEE International Conference on Communications, ICC 2002, vol. 2, pp. 1199–1203. IEEE (2002)

6. Ding, L., Goubran, R.A.: Speech quality prediction in VoIP using the extended e-model. In: IEEE Global Telecommunications Conference, GLOBECOM 2003, vol. 7, pp. 3974–3978. IEEE (2003)
7. Nguyen, D.-H., Nguyen, H.: A dynamic rate adaptation algorithm using WB E-model for voice traffic over LTE network. In: 2016 9th IFIP Wireless Days (WD). IEEE (2016)
8. 3rd Generation Partnership Project (3GPP): http://www.3gpp.org
9. Ghalut, T., Larijani, H.: Non-intrusive method for video quality prediction over LTE using random neural networks (RNN). In: 2014 9th International Symposium on Communication Systems, Networks & Digital Signal Processing (CSNDSP), pp. 519–524. IEEE (2014)
10. Hyun, J., Li, J., Im, C., Yoo, J.H., Hong, J.W.K.: A volte traffic classification method in LTE network. In: 2014 16th Asia-Pacific Network Operations and Management Symposium (APNOMS), pp. 1–6. IEEE (2014)
11. ITU-T: ITU-T recommendation P.800: methods for subjective determination of transmission quality. Technical report, International Telecommunication Union, August 1996
12. ITU-T: ITU-T recommendation G.107: the e-model, a computational model for use in transmission planning. Technical report, International Telecommunication Union, December 1998
13. ITU-T: ITU-T recommendation P.862: perceptual evaluation of speech quality (PESQ): an objective method for end-to-end speech quality assessment of narrow-band telephone networks and speech codecs. Technical report, International Telecommunication Union, February 2001
14. Larijani, H., Radhakrishnan, K.: Voice quality in VoIP networks based on random neural networks. In: 2010 Ninth International Conference on Networks (ICN), pp. 89–92. IEEE (2010)
15. The Mathworks, Inc., Natick, Massachusetts: MATLAB version 8.5.0.197613 (R2015a) (2015)
16. Meddahi, A., Afifi, H., Zeghlache, D.: Packet-e-model: e-model for wireless VoIP quality evaluation. In: 14th IEEE Proceedings on Personal, Indoor and Mobile Radio Communications, PIMRC 2003, vol. 3, pp. 2421–2425. IEEE (2003)
17. Mushtaq, M.S., Augustin, B., Mellouk, A.: Qoe-based LTE downlink scheduler for VoIP. In: 2014 IEEE Wireless Communications and Networking Conference (WCNC), pp. 2190–2195. IEEE (2014)
18. Olariu, C., Foghlu, M.O., Perry, P., Murphy, L.: VoIP quality monitoring in LTE femtocells. In: 2011 IFIP/IEEE International Symposium on Integrated Network Management (IM), pp. 501–508. IEEE (2011)
19. Piro, G., Grieco, L.A., Boggia, G., Capozzi, F., Camarda, P.: Simulating LTE cellular systems: an open-source framework. IEEE Trans. Veh. Technol. **60**(2), 498–513 (2011)
20. Radhakrishnan, K., Larijani, H., Buggy, T.: A non-intrusive method to assess voice quality over internet. In: 2010 International Symposium on Performance Evaluation of Computer and Telecommunication Systems (SPECTS), pp. 380–386. IEEE (2010)
21. Reisslein, M., Karam, L., Seeling, P.: H. 264/AVC and SVC Video Trace Library: A Quick Reference Guide (2009). http://trace.eas.asu.edu
22. Takahashi, A., Kurashima, A., Yoshino, H.: Objective assessment methodology for estimating conversational quality in VoIP. IEEE Trans. Audio Speech Lang. Process. **14**(6), 1984–1993 (2006)

A Modular Secure Framework
Based on SDMN for Mobile Core Cloud

Karim Zkik[✉], Tarik Tachihante, Ghizlane Orhanou, and Said El Hajji

Laboratory of Mathematics, Computing and Applications, Faculty of Sciences,
University of Mohammed V in Rabat, Rabat, Morocco
karim.zkik@gmail.com, tariktachihante@gmail.com,
{orhanou,elhajji}@fsr.ac.ma

Abstract. During these last few years, mobile data traffic has been
strongly growing while the voice traffic decreases (fixed and mobile), This
new reality pushes operators to invest in a new next generation of mobile
network (Mobile Cloud Computing, 5G) to enhance their competitivity
and provide more innovative marketing products and services to the end
user. But they were facing a major obstacle: the computer networks
management is too complex and difficult. The research community proposes Software Defined Mobile Networks (SDMNs) as a solution to provide more flexibility and to ease the management of the next-generation
mobile networks especially Mobile Cloud Computing (MCC). This new
solution offers a huge advantage to the mobile operators and enables
innovation through network programmability. SDMN provides several
benefits including, network and service customized, improved operations
and better performance, but there are some security issues that need to
be taken care of. This paper describes the emergence of SDMN as an
important new networking technology, discusses the different problems
related to security issue. We propose a framework to secure the different
levels in SDMN architecture with an implementation of our framework.
A simulation has been done of some main threats such as DDoS attack
and malware infection.

1 Introduction

Mobile Cloud Computing refers to availability of cloud computing services in
a mobile environment. MCC is composed of three main components: wireless
communication channel, mobile devices, and cloud. MCC offer many advantages to mobile users such as breaking through the mobile hardware storage and
calculation limits, and On-demand self-service share and access to data. The
mobile cloud computing has become increasingly popular and become a focus of
research. In 2015, mobile traffic reached about 30% of the total Internet traffic
[1], and a growth of 66% in this traffic is estimated by 2017 [2].

The exponential use of the MCC services causes several problems in terms
of configuration and management, because the traditional mobile networks are
complex and hard to manage and we must configure each individual network

S. Boumerdassi et al. (Eds.): MSPN 2016, LNCS 10026, pp. 137–152, 2016.
DOI: 10.1007/978-3-319-50463-6_12

device separately. To avoid these problems, researches has focused to find an automatic reconfiguration and response system, able to control the entire network, offering agility, simplicity, provability, innovation and new abstraction. Software-Defined Mobile Networking (SDMN) proposed as an emerging networking paradigm, derived from Software-Defined Networking (SDN), to change the limitations of current mobile network infrastructures. SDN offers the separation of the control and data planes, and is characterized by five fundamental traits: plane separation, a simplified device, centralized control, network automation, virtualization and programmability, and openness [3,4]. The controller as a brain of the network, it will be responsible for maintaining the whole network topology, implements policy decisions, regarding routing, forwarding, redirecting and control all the SDN devices, and provides a northbound API for mobile applications and southbound API for protocol to communicate with network elements. The OpenFlow Protocol is the most important component of SDN and SDMN, since it ensures communication between the control plane and the data plane. An OpenFlow switch maintains a forwarding table (flow table) that contains a list of prioritized rules (flows).

Despite of the problems related to the configuration and management of networks, MCC has several problems related to data and network security [5,6]. These problems gets worse while using SDMN, because the entire network is managed by a single controller which is consider as a failure point. Using SDN is based on Network Function Virtualization (NFV) [7] which reduces running costs and increases system flexibility, but it causes some additional security problems because classical security devices can not protect and monitor the components of virtualized environments. In this article we propose an architecture that is designed to secure the different Plan of SDMN architecture. For that we will draw up the main security problems of the data plane and the control plane and we propose some appropriate security measures for each case. It also proposed an implementation of these security measures by using virtualization and a security analysis of our framework. In this security analysis we demonstrate that our architecture can offer authentication integrity and confidentiality of circulating flow through SDMN networks.

The paper is organized as follows: In Sect. 2, we present the SDMN architecture and problematic definition. In Sect. 3, we present in details our proposed model and we describe the different element of our security framework. In Sect. 4, we present an implementation of our framework with some real attack situations and we provide a performance evaluation results. Finally, in Sect. 5, we conclude the paper.

2   SDMN Presentation

2.1   SDMN Architecture

Recently the mobile traffic increased drastically while providing multiple services such as VoIP (Voice over IP), video streaming and virtual network services [8]. Mobile Cloud Computing (MCC) has played a very important role in this rapid development of mobile networks, because it has provided platforms for out-sourcing data and several service models for users (SaaS, IaaS and PaaS). The exponential use of MCC causes limitation on radio bandwidth resources and inflexibility in network equipment. So, the Mobile and Cloud service providers start thinking about solutions that facilitate the management of flows and mobile networks.

The mobile networks continues to evolve and many technologies and services related to mobile and mobile cloud are born every day, and operators invest more and more in new next generation mobile network to enhance performance and scalability of telecommunication networks. So, several studies have been conducted to find a solution that will allow to manage all this huge architecture. Software-Defined Mobile Network (SDMN) is the most accomplished solution that can offer many possibilities in term of flow management. The concept of the Software-Defined Mobile Network (SDMN) is introduced as an extension of SDN to combine mobile network specific functionalities and to adapt SDN to users and operators needs in terms of agility, flexibility and bandwidth resources. Figure 1 illustrates a traditional SDMN architecture [9]. Software Define mobile Networking (SDMN)is a centralized architecture which permits to manage the flow from controller, while using OpenFlow protocol (OF), to facilitate communication between the control plane and data plane. SDMN offers more portability and elasticity to mobile network but this centralized system opens up new security challenges and threats.

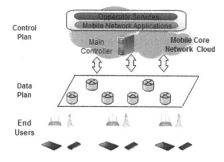

Fig. 1. Traditional SDMN architecture

2.2 SDMN Security Issues

Today, we need new networks that are flexible, programmable, and manageable to follow the emergence of new technologies such as 5G, MCC or cloud computing. Using SDN architecture is one of big step in that direction but it poses a new security challenges and threats:

1-SDMN architecture is centralized around a control plan and especially the controller, which is responsible for controlling network devices and to make decisions. So we may consider it, as a single point of failure and any problem occurring in the controller can impact the functioning of our entire network [10].

2-The Open Flow [11] protocol is considered as the southbound communication protocol between the data and control Plan, but its vulnerable to SSL attacks [12] which could deteriorate the performance and cause a serious security problem in terms of confidentiality and integrity.

3-The network intelligence that is normally located within each network device will be logically centralized in controller. This new concept makes the whole data plan and network devices as a simple packet forwarding without a real intelligent or security mechanism to protect their self from potential threats.

4-The overall mobile networking industry and operators are facing some serious security problems, because the mobile devices are much easy to be affected by malware and virus due to OS and applications vulnerabilities. An attacker can use those vulnerabilities to usurp the users' identities and steal sensitive information or infect the whole network.

2.3   Related Work

To address these security problems several studies have been conducted [13] and several architectures have been made Madhusanka Liyanage and et al. [14], propose a secure communication channel using IPSec tunneling and a security gateway to protect the flow in mobile networks. Paulo Fonseca and et al. [15] propose a replication component (CPRecovery) for resilient OpenFlow by using a primary and secondary controller to ensure the high availability of the network. Rodrigo Braga and et al. [16] propose some detection techniques using NOX controller to protect the flow and detect DDoS attacks in the network. Seugwon Shin and et al. [17] propose a modular security services for SDN using FRESCO to defend the network from detected threats. Hongxin Hu and et al. [18] propose a firewall named FLOWGUARD for Software-Defined Networks, this firewall permits to detect and resolve violations through the network.

All this researches have led to very good results, and it can be considered as a solid basis for building a more robust security platform that secures all the SDN layers and all network component, and more specifically relates to mobile network security and MCC environment.

3   Proposed Security Framework for MCC Using SDMN

3.1   General Security Scheme

The goal of our work is to demonstrate the benefits of integrating the SDN in the mobile network and in MCC environment, and offering a security solution dedicated to SDMN, which secures all layers, equipment, devices and users of our network. Figure 2 describes the general proposed SDMN architecture. The purpose of our architecture is to secure the whole SDMN architecture. We added some security components in each layers, some security modules in SDN and SDMN devices and some secure communication protocols to detect and resolve infections and to enhance security. To describe our framework we will make an analysis of each layer, list the existing breaches and vulnerabilities, and present the developed solution.

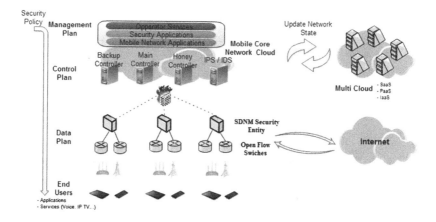

Fig. 2. SDMN proposed architecture

3.2   Control Plan

Control Plan maintains the global view of the network, forwards updates on open flow tables to data plan, and make decision regarding routing and data flow. SDMN is a centralized system and the controller is the heart of this system, making it a failure point. So, utilizing a centralized approach in order to provide network services can lead to serious problems:

– Any problem in the server in which the controller process is running (Shut-down, infection or overload of server), can lead to serious damage in the entire network.
– It is also likely that a management application entering into an infinite loop will then overload and block the network [19].
– Or even, an attack from another network can lead to take a possession of the controller. So, the attacker can update and redirect traffics as he pleases.

To avoid all these problems we propose two security measures to protect our control plan. So, firstly we propose a replication mechanism to ensure the hight availability of our network, and secondly we propose an intrusion detection mechanism based on a SDMN firewall, an IPS/IDS server and a honey controller to detect stop and manage any sort of attacks or infections.

Replication mechanism: Replication must exist to guarantee the high availability of our network. So, in case of failure of the main controller another controller can takes over. On the basis of works of Paulo Fonseca and et al. [15] we made an appropriate replication mechanism to our architecture. Figure 3 shows the replication mechanism that we have propose.

Our first priority is to detect any abnormal behavior. For that we place at the control plane an IDS server that will allow us to monitor the controller in real time and detect any kind of intrusion. The security and policy update are

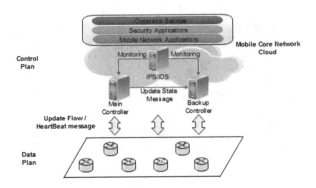

Fig. 3. Replication in control plan layer

periodic and made from the applications of security located in the management plan. The replication mechanism works as follows:

The entire communication between the control plane and the data plane are orchestrated by the main controller that receive requests from the open flow switches and sends updated of our open flow tables. Communications are maintained as the heartbeat messages are maintained and as long as no security problem is detected at the level of the established security devices through our architecture. To ensure the high availability of our architecture, we implemented a second controller to takes over our main controller in case of failure. For this, the backup controller periodically receives updates of network status from the main controller. In case of failure of the main controller, the backup controller automatically takes the lead.

Intrusion detection and resolution mechanism: Several attacks aim to send suspicious packets to the Control Plan in order to infect the controller. To avoid this problem, we have implemented an Intrusion detection and resolution mechanism with an adapted firewall for our SDMN architecture on the basis of works of Hongxin Hu and et al. [18] which propose FlowGuard firewall for SDN architecture. Figure 4 shows the security measures that we have established to secure the data plane from infections and security breaches.

When a new flow comes it is immediately analyzed by our SDMN firewall. To detect security violations the SDMN Firewall check the packet headers to verify the validity of the source and destination addresses. Since the use of the controller allows us to have a global view of the network, the firewall SDMN knows in advance the number of hop that a packet should do before reaching its destination. So if the number of hops made by a packet is different, the firewall rejects immediately the packet and sends a report to the Control Plan for analysis. SDMN firewall is also responsible for checking rules and flow policy, these policy update are made periodically from the Management Plan. The flow verification algorithm is shown in Fig. 5. In case of flooding attack, SDMN firewall blocks the flow when exceeds a certain level. A security violation log is sent immediately to the controller, so it can take a decision in this regard.

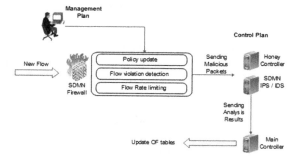

Fig. 4. Intrusion detection and resolution mechanism

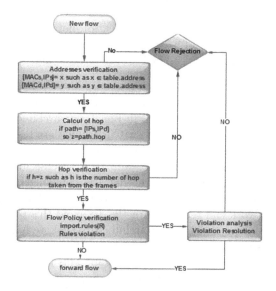

Fig. 5. Intrusion detection and resolution algorithm

SDMN firewall is connected to an IDS, which analyzes the new flow in order to detect possible infection. The virus database is updated from the management plan and from our mobile cloud security applications. In case of infection, the packets are sent to the honey controller. The honey controller gives attackers the illusion of having succeeded their infection. So, the IDS server can analyze the corrupted packets and know the source and the purpose of the attack. An infection log is then sent to the controller, so it can take a decision in this regard.

3.3   Data Plan

Data plane corresponds to the networking devices, which are responsible for forwarding traffic according to the decisions made by the control plane. In the old architecture the development and the deployment of new networking features is very hard because it is imperative to configure each device separately, which is extremely time consuming and increases the risk of errors and failure in the system. The separation of data plan and control plan in SDN offers a solution to all these problems, and it offers huge possibilities in terms of network management, but this separation causes loss of intelligence at the level of data plane and creates several security vulnerabilities:

- Controller usurpation: In SDN or SDMN architecture security depends on controller security. So, an attacker can send flow and incorrect rules by impersonating a controller, to deceive switches and redirect flow to undesired destinations.
- Flooding attacks: Flow tables of Open Flow switches can store a limited number of flow rules, which can causes deny of services in case of a flooding attacks. [20]
- Communication weaknesses: The communication between open flow switches in SDN is made by using TLS/SSL communications, which makes it vulnerable to TLS/SSL weaknesses, TCP-Level attacks, and Man-in-the middle attack. [21]

It is proposes a set of measures against all this threats to secure our architecture and especially the Data Plan. As shown in Fig. 6, our architecture is composed of two essential elements: (1) open flow switches which are responsible for redirecting traffic to the cloud, and (2) the SDMN security entity which responsible to secure and monitor in and out flows, and to ensure secure communication between Control and Data Plan. Given that the controller is responsible for managing the flow of our network, it is considered as a failure point and any infection on the network can reach it directly. To avoid this problem, we added a number of security equipment to our architecture that we have named SDMN security entity. This equipment is primarily responsible for securing the circulating flow between the controller and the data plan. This measure will allow us to separate in some way switches from controller. So, switches will no longer have any visibility on the controller, which will increase the level of security.

The SDMN security entities will be distributed throughout the network, and each device will be responsible to manage and monitor a group of switches. So, in case of attack or failure, the infection will be neutralized in a single group of switches at most, even more the controller will be safe from any attacks launched from the Data Plan layer.

The SDMN security entities are composed of several modules: (1) the first module is responsible for verifying packets header, the port number and the IP and Mac address, (2) the second module is responsible for further verification and inspect the application layer to detect any suspicious codes, (3) the third module verifies if the flow rules are respected, (4) the fourth module analyzes the flow to detect any trace of malware that could compromise or infect the network. The malware database is periodically updated from the Control Plan. When a module detects an abnormality or abnormal behavior, it interrupts the communication outcome of a source of the problem and sends a report to the control plan for treatment.

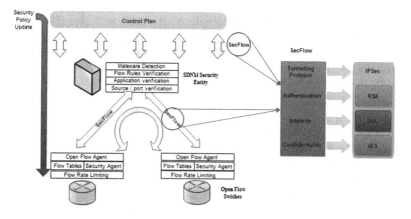

Fig. 6. Data plan layer

The OpenFlow switches are bound to end users and Control Plan, and they are responsible for redirecting traffic circulating between users and the Mobil cloud. Ensuring a security of these switches is a key task to prevent attacks and penetration attempts on our network. So in order to do this, we added two security modules to our switches: (1) a flow rate limiting who is responsible for stopping the flow in case of flooding attack. You can configure the module to close the switch ports as soon as the rate exceeds certain threshold, (2) a security agent who is responsible to verify signatures and authentication requests from the Control plan, enabling a safely update of flow tables and preventing spoofing attacks.

To provide a higher level of security, it is imperative to secure the communications between the various entities of the data plane and the communications between the control and data plan. So, we have provided a communication protocol named SecFlow based on ipsec tunneling protocol which ensures the authentication, integrity and confidentiality. Figure 6 shows also the different components of our communication protocol "SecFlow" and the different crypto system used in the design of each component.

4 Simulation Result and Proof of Concept

4.1 Environment

The proposed testbed includes network SDMN devices which are SDN controller, Backup controller, honey controller, IPS/IDS server, SDMN security entity, SDN switch and mobile devices. We used experimentation setups for our Proof of Concept hierarchy network model and we devise our topology into 3 domains, each network domain A, B and C is managed by a local SMDN security entity and all of them are connected directly to the controller and backup controller. This setup is representative from a typical operator bear network where several sites (edge the topologies is separated into several sub-networks).

We use a PC Server (HP DL380G6 Xeon quad-core E5504 2.00 GHz/ 4-core/4 MB/80 W/Memory 24 GB). Under the management of an Open source virtualization platform XEN server [22], we installed the following guest operating systems: OpenDayLight Controller Lithium on Linux operating system Ubuntu 12.10 64 bits with 8 GB of RAM; Mininet Emulator [23] version 2.2.1 on Linux operating system Ubuntu 10.12 64 bits with 1 GB of RAM.

We created 4 Virtual Machines (VM) on XEN server, the first one (1) to perform the SDMN security entity that comprise 4 modules (antivirus, flow rate limiting, packets filtering using IP/Mac address, port number verification) and the second (2) Backup Controller the third (3) an IPS/IDS server and a Honey Controller and finally (4) a virtual machine to simulate our main controller. All VMs are connected to an Open VSwitch interface [24].

We used some sniffing tools and monitoring software to calculate and monitor flows on our network and we used IPv6 addresses to configure all devices.

4.2   Study of Some SDMN Infection Cases

In this section we will simulate two types of attack on SDMN networks, describe the different steps of these attacks, and provide the different detection, prevention and resolution mechanisms of these attacks. It is chosen to simulate a malware infection and a DDoS attack to test the different established security mechanisms and demonstrate the robustness of our architecture.

Scenario I: Malware infection. SDN is based on a centralized architecture on the controller, which means that the various entities of our networks lost in some way their intelligence, which makes them more vulnerable to attacks and infections. Today's malware are much developed and become more sophisticated. Some malwares have become polymorphic, which mean they can change in nature which makes them difficult to detect. Malwares are also very persistent and very dangerous and they can cause considerable damage in our network. To protect our network against malware infections we will proceed as follows:

– First we'll make sure to detect any infiltration on our networks, monitor all communications and circulating flows, and remove any infection detected in our network.
– Secondly we will try to protect our controller against any kind of malicious infection by redirecting flow to the honey controller.

As shown in Fig. 7, an attacker has the ability to infect mobile and OF Switches or directly infect the Control Plan. The detection at the level of access and data plane is made at the level of SDMN security entity and the detection at the level of control plane is made from the IPS/IDS. After having detected the infection, malicious flow is redirected to a honey controller, and the attacker will seem to have succeeded the attack. This will allow us to analyze the malicious flow, know the purpose of the attack, and in several cases know the identity of the

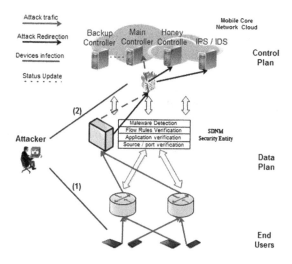

Fig. 7. Malware detection and analysis mechanism

Table 1. Malware infection analysis

		Flux analysis (Packet/s)	Detection (s)	Infection analysis (s)	Action/ Priority	Latency (us)	Detection (s)
SDMN security entity	Normal flow	123	–	–	Allow/1	4.2	137.4
	Infected flow	142	12.15	24.6	Forward to IPS/IDS		
Firewall	Normal flow	215	–	–	Allow/1	3.4	106.7
	Infected flow	238	7.12	14.7	Forward to IPS/IDS		
Honey controller	Normal flow	196	–	–	Allow/1	3.6	–
	Infected flow	134		428.3	Remove/3		
IPS/IDS	Normal flow	156	–	–	Allow/1	2.5	–
	Infected flow	174	5.71	428.3	Deny/3		

attacker. Table 1 shows the different steps of detecting and analyzing malicious flows, giving the security mechanism that we have set in our architecture.

Scenario II: DDoS attack. A DoS attack and its distributed version, Distributed Denial-of-Service attacks (DDoS) are one of the most dangerous security problems in IT networks. This attack is a flooding attack type, and it aims to make network unavailable. In what follows we will simulate a DDoS attack, and show the various measures and security mechanism that will permit us to detect and stop this type of attack.

A DDoS attack can be launched by infecting the OpenFlow Switches and then create a botnet network. This botnet network will flood the network with packets and thus launch a DDoS attack. To prevent this type of attack we split our network into multiple sub networks, in our implementation we used 3 virtual subnets by using MININET, and each Subnet is linked to an SDMN security entity server. This measure allows dividing the network into several areas, and in case of flooding or DDoS attack we can neutralize the attack in one area rather than risk infecting the whole network. Figure 8 shows the progress of the DDoS attack in one area. This attack aims to disrupt the functioning of all equipments in our network, especially saturating and dropping the controller.

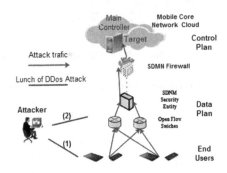

Fig. 8. DDoS detection and resolution mechanism

We added an additional module in each OpenFlow switches and each SDMN security entity. This module is a flow rate limiting and it is responsible to stop the flow overflowing above a certain rate. Once the violation detected, the module automatically closes the switch and the server ports, and send a violation report to the controller to take the appropriate decision. The flow rate limiting module calculates the average number of packets allowed for each port (PNL) according to the number of inflow (FN) and the maximum number of packets allowed by flow (PN) using the following equation:

$$PNL = (PN * FN) \tag{1}$$

Table 2 describes the results of the attacks following the scenario and the layer in which it was launched, and also according to type of DDoS attacks employed.

4.3   Security Analysis

We designed a secure architecture for SDMN networks that covers all the layers. The purpose of this architecture is to propose countermeasures and solutions against all threats that can reach one or several layers of SDMN networks. Table 3 illustrates the critical components for SDN and SDMN networks and compares

Table 2. DDos attack analysis

	DDoS from access plan			DDoS from data plan		
DDoS attack types	ICMP	UDP	TCP/SYN	ICMP	UDP	TCP/SYN
OpenFlow switch 1 (packets)	3049	2118	2879	1156	945	2213
OpenFlow switch 2 (packets)	2360	1425	3078	2672	1102	2967
SDMN security entity (packets)	1053	986	1210	3156	4652	6249
Detection times(s)	2.54	1.68	3.04	6.14	7.06	8.93
CPU times(s)	260	158	294	147	123	224

their use between the existing architecture and our architecture. The flow in the network is guided by a number of rules; these rules are checked by our SDMN firewall. Figure 9 shows the steps of analysis, detection and resolution of violations flow at the level of our SDMN firewall.

Our architecture has also a replication mechanism that ensures the high availability in case of failure of the main controller. It also features an IPS/IDS and a honey controller responsible for analyzing and stoping infected packets. The reports and logs of violation are directly sent to the controller to make the appropriate decision and update open flow tables. Figure 10 shows the replication mechanism in case of failure of the main controller and illustrates the simulation results of DDoS attacks.

Table 3. critical security components in SDMN

	M. Liyanage and et al. [14]	Hongxin Hu and et al. [18]	P. Fonseca and et al. [15]	R. Braga and et al. [16]	S. Shin and et al. [17]	Our architecture
Mobile Cloud protection	Yes	No	No	No	No	Yes
IPS/IDS	No	No	No	No	Yes	Yes
Firewall	No	Yes	No	No	No	Yes
Network resilience	No	No	Yes	No	No	Yes
Network monitoring	Yes	No	Yes	Yes	Yes	Yes
Flow rules verification	No	Yes	No	No	Yes	Yes
Configuration verification	Yes	Yes	No	Yes	Yes	Yes
Controller availability	Yes	No	Yes	Yes	Yes	Yes

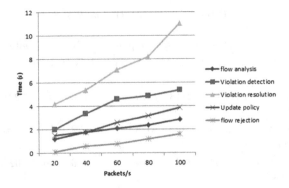

Fig. 9. Flow rules analysis

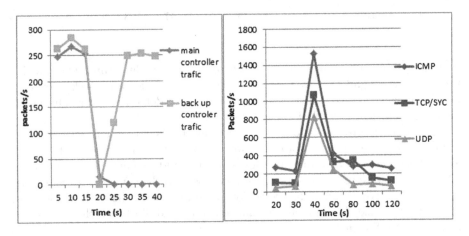

Fig. 10. Replication and DDoS detection and resolution mechanism

5   Conclusion

The SDN and SDMN networks are becoming increasingly popular, and they will soon replace the existing traditional network architectures as they offer enormous facilities in terms of management of networks and it provides agility and flexibility. The use of SDN networks in mobile networks and in mobile cloud environment offers huge benefits for mobile operators and cloud services concerning the management of flows, but there are several security problems disrupts the use and the deployment of such networks. Our architecture allows us to offer more security to flow in all layers of SDMN networks. At the data plan we have added security modules and access verification in all OF Switches, we have developed server which we named SDMN security entity, which is responsible to analyze the incoming and outgoing flows and to make a separation between the data plan and control plan and we have established a communication protocol that we named SecFlow which enables secure communication between the various entities of the data plan. In the control plan, we have developed an SDMN firewall, which is responsible for analyzing the incoming and outgoing flows of control plan and detect violations of flow policy, we have developed a replication mechanism using a backup controller that takes over in case of failure of the main controller and we have developed a detection and resolution mechanism in case of infection on the network using an IPS/IDS server and honey controller.

Concerning the simulation part we have created a virtual SDMN architecture using MININET, OpenDay Light and V-Switch, and we simulated a DDoS attack and malware infection. We established also a security analysis which sets the various security aspects that we have processed.

Concerning our future work, it is expected to deploy our architecture on a real network, enhancing the security protocols, and minimize latency. It is also interested in distributed SDN and SDMN architectures because there is an interconnection between multiple controllers, and we would like to provide adequate framework architecture for this type of networks. It also plans to work on the OpenFlow protocol and incorporate it some security mechanisms and to generalize it in all existing SDN and SDMN networks.

References

1. Ericsson Mobility Report (2015)
2. Cisco Visual Networking Index: Global Mobile Data Traffic Forecast Update 20152020 (2016)
3. Kreutz, D., Ramos, F.M.V., Verissimo, P., Rothenberg, C.E., Azodolmolky, S., Uhlig, S.: Software-defined networking: a comprehensive survey. Proc. IEEE **103**(1), 14–76 (2015)
4. Masoudi, R., Ghaffari, A.: Software defined networks: a survey. J. Netw. Comput. Appl. **67**, 1–25 (2016)
5. Ryan, M.D.: Cloud computing security: the scientific challenge, and a survey of solutions. J. Syst. Soft. **86**, 2263–2268 (2013)
6. Fernando, N., Loke, S.W., Rahayu, W.: Mobile cloud computing: a survey. Fut. Gener. Comput. Syst. **29**, 84106 (2013)
7. Bellavistaa, P., Callegatia, F., Cerronia, W., Contolic, C., Corradia, A., Foschinia, L., Pernafinia, A., Santandrea, G.: Virtual network function embedding in real cloud environments. Comput. Netw. **93**, 506517 (2015). Part 3
8. Naboulsi, D., Fiore, M., Ribot, S., Stanica, R.: Large-scale mobile traffic analysis: a survey. IEEE Commun. Surv. Tutorials **18**(1), 124–161 (2015). IEEE Communications Society, Institute of Electrical and Electronics Engineers
9. Pentikousis, K., Wang, Y., Weihua, H.: Huawei Technologies, MobileFlow, Toward SoftwareDefined Mobile Networks (2012)
10. Yao, G., Bi, J., Guo, L.: On the cascading failures of multicontrollers in software defined networks. In: 2013 21st IEEE International Conference on Network Protocols (ICNP) (2013)
11. McKeown, N., Anderson, T., Balakrishnan, H., Parulkar, G., Peterson, L., Rexford, J., Shenker, S., Turner, J.: OpenFlow: enabling innovation in campus networks. ACM SIGCOMM Comput. Commun. Rev. **38**(2), 6974 (2008)

12. Liyanage, M., Ylianttila, M., Gurtov, A.: Securing the control channel of software-defined mobile networks. In: 2014 IEEE 15th International Symposium on a World of Wireless, Mobile and Multimedia Networks (WoWMoM), p. 16. IEEE (2014)
13. Ahmad, I., Namaly, S., Ylianttilaz, M., Gurtov, A.: Security in software defined networks: a survey. IEEE Commun. Surv. Tutorials **17**(4), 2317–2346 (2015)
14. Liyanage, M., Ahmed, I., Ylianttila, M., Santos, J.L., Kantola, R., Perez, O.L., Itzazelaia, M.U., de Oca, E.M., Valtierra, A. and Jimenez, C: Security for future software defined mobile networks. In: 2015 9th International Conference on Next Generation Mobile Applications, Services and Technologies (2015)
15. Fonseca, P., Bennesby, R., Mota, E., Passito, A.: A replication component for resilient openflow-based networking. In: IEEE Network Operations and Management Symposium (NOMS): Mini-Conference (2012)
16. Braga, R., Mota, E., Passito, A.: Lightweight DDoS flooding attack detection using NOX/OpenFlow. In: 35th Annual IEEE Conference on Local Computer Networks (2010)
17. Shin, S., Porras, P., Yegneswaran, V., Fong, M., Guofei, G., Tyson, M.: FRESCO: modular composable security services for software-defined networks. In: ISOC Network and Distributed System Security Symposium (2013)
18. Hongxin, H., Han, W., Ahn, G.-J., Zhao, Z.: FLOWGUARD: building robust firewalls for software-defined networks. In: Proceedings of the Third Workshop on Hot Topics in Software Defined Networking, HotSDN 2014, pp. 97–102 (2014)
19. Luo, S., Hongfang, Y., Li, L.: Practical flow table aggregation in SDN. Comput. Netw. **92**, 7288 (2015). Part 1
20. Jarraya, Y., Madi, T., Debbabi, M.: A survey and a layered taxonomy of software-defined networking. IEEE Commun. Surv. Tutorials **16**(4), 1955–1980 (2014)
21. Benton, K., Camp, L.J., Small, C.: OpenFlow vulnerability assessment. In: Proceedings of the Second ACM SIGCOMM Workshop on Hot Topics in Software Defined Networking, HotSDN 2013, p. 151152. ACM (2013)
22. XenServer 6.x Best Practices. Dell Compellent Storage Center (2013)
23. Introduction to Mininet. https://github.com/mininet/mininet/wiki/Introduction-to-Mininet
24. Open vSwitch Configuration Guide. Configuration Guide for the OVS PICA8 Switch, 1st edn. (2011)

Cloud Security Quantitative Assessment Based on Mobile Agent and Web Service Interaction

Abir Khaldi[✉], Kamel Karoui, and Henda Ben Ghezala

RIADI Laboratory, ENSI, University of Manouba, Manouba, Tunisia
abirakaldi@yahoo.fr,
kamel.karoui@insat.rnu.tn,
hhbg.hhbg@gmail.com

Abstract. Cloud security is very challenging and is becoming a research hot topic. Thus, the adoption of the security assessment would be the key to evaluate and to enhance the cloud security level. The security assessment can be quantitative or qualitative. This paper proposes a cloud security quantitative assessment (*CSQA*) model. This proposed model evaluates the security of any cloud service (*XaaS*) exposed to attacks and vulnerabilities affecting its quality and specially its availability. It is based on mobile agent and web service interaction framework.

Keywords: Cloud security · Security assessment · Mobile agent · Web service

1 Introduction

Based on the cloud adoption practices and priorities survey report in January 2015, cloud security projects were the leading IT project in 2014 [1]. As illustrated in Fig. 1, 75% of companies qualified cloud security projects as important or even very important, in addition to eclipsing intrusion prevention (74%) besides firewalls and proxies (65%).

This stresses on the necessity of continuously improving cloud security and ensuring its sustainability towards customers varying from enterprises to end users.

We propose a three-phase flow hierarchical model for cloud security quantitative assessment (*CSQA*). We first present the situation perception concept this is followed by the situation evaluation concept presentation and finally the situation prediction that covers all cloud services.

In order to apply this model, we design a *CSQA* framework based on mobile agent and web service interaction.

This paper is organized as follows: Sect. 2 discusses the related works. Section 3 proposes the *CSQA* model. Section 4 focus on the security situation evaluation for all the cloud services. Section 5 describes the *CSQA* framework based on mobile agent and web service interaction. Section 6 is a case study and Sect. 7 evaluates the proposed framework performance. Finally, Sect. 8 gives a conclusion for the paper.

© Springer International Publishing AG 2016
S. Boumerdassi et al. (Eds.): MSPN 2016, LNCS 10026, pp. 153–167, 2016.
DOI: 10.1007/978-3-319-50463-6_13

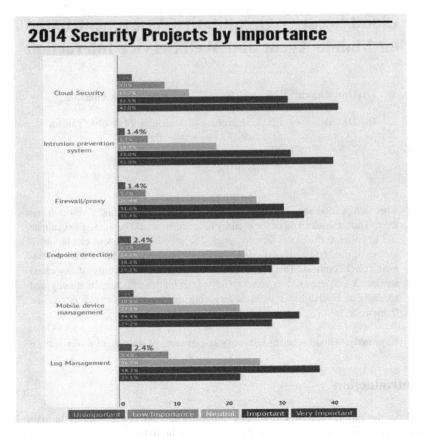

Fig. 1. 2014 Security projects by importance [1]

2 Related Work

Endsley [2] gave the first situation awareness definition since 1988. Initially, this concept was used for aviation and military purposes to evaluate the security situation awareness. Deriving from global situation awareness, the network security situation awareness (NSSA) appeared as a very interesting research field aiming to describe the status of network equipments, network behaviors and user actions.

Bass [3] first presented the concept of the Network situation awareness (NetSA) in 1999. The concept was based on air traffic control (ATC) situation awareness, and developed a mature theory and technology from ATC situation awareness.

Many researchers [4–7] have focused on NSSA emphasizing three-phases flow:

- **Phase 1.** Situation Perception: analyzes and categorizes original data. This module prepares for situation evaluation.
- **Phase 2.** Situation Evaluation: data outputs from Phase 1 is analyzed within this phase using a precise mathematical model. It provides a comprehensive and quantitative description of current situation.

- **Phase 3**. Situation Prediction: it displays actual situation map and forecasts the future situation.

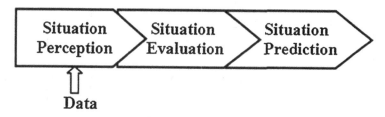

Fig. 2. The *NSSA* conceptual models

Several researches [8–11] deployed intrusion detection and prevention systems in order to evaluate cloud environment security. Their works provided a framework to collect attacks in order to evaluate qualitatively the security level in the cloud.

In [12], authors focused on the vulnerabilities evaluation. They proposed the VULCAN model to evaluate the vulnerabilities in the cloud platform. They used an ontology knowledge base (OKB) to develop an automatic process capable of evaluating cloud security level within a global cloud architecture. We can consider this VULCAN model as a vulnerability detection system.

Authors in [13] developed the risk assessment to evaluate security in the cloud. they proposed a *QUIRC* model based on the risk assessment definition. the definition combined the threat probability and its severity in relation to its impact. The proposed *QUIRC* model used the Delphi method to collect required data for risk assessment. A risk knowledge database is generated over time helping to construct an objective database for the risk evaluation. This model proposed to assess security based on cloud risks for all its platform.

The cloud provider in [14] conducted a risk assessment with the participation of its clients. In fact, this qualitative assessment can be more realistic and more precise helping provider to enhance cloud security level due to the customer feedbacks.

In [15], they proposed a risk assessment within cloud provider in order to choose the most secure one. It is a generic evaluation covering the global provider services.

In Table 1, we discuss the cited works [8–15] based on the assessment method (qualitative or quantitative), the assets to evaluate (cloud platform or service platform), the security level of the assessment platform and the threats to detect (attack or vulnerability).

After related works discussion, we propose a Cloud Security Quantitative Assessment (*CSQA*) to evaluate the security for any cloud service. The proposed *CSQA* framework should be secure and able to detect attacks and vulnerabilities for all cloud assets. Next, we will detail our *CSQA* model.

Table 1. A related work comparative table

Works	Assessment		Cloud platform	Cloud service	Secure Platform	Detection	
	Qualitative	Quantitative				attack	Vulnerability
[8] [9] [10] [11]	Yes	No	Yes	No	No	Yes	No
[12]	Yes	No	Yes	No	No	No	Yes
[13]	No	Yes	Yes	No	No	Yes	No
[14]	Yes	No	Yes	No	No	Yes	No
[15]	No	Yes	Yes	No	No	Yes	No

3   The Hierarchical CSQA MODEL

In the traditional networks, CHEN's et al. [4] proposed an hierarchical threat evaluation model. According to this model (Fig. 3), the traditional networks could be divided, based on scale and hierarchy, into local networks, hosts, services and threats.

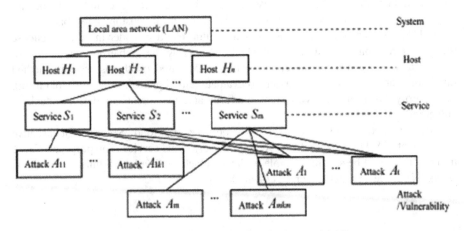

Fig. 3. Hierarchical threats evaluation model [4]

This model supposes that the index of network system level security situation is the weighted sum of all the host level index. Where this last one refers to the number of different important services simultaneously suffering threats at a given moment.

We noticed that the same hierarchical threats evaluation model can be applied to the cloud services. NIST [16] defined three models of cloud services which are:

- Software as a Service (*SaaS*): The capability provided to the consumer is to use the provider's applications running on a cloud infrastructure.
- Platform as a Service (*PaaS*): The capability provided to the consumer is to deploy onto the cloud infrastructure consumer - created or acquired applications created using programming languages, libraries, services, and tools supported by the provider.
- Infrastructure as a Service (*IaaS*): The capability provided to the consumer is to provision processing, storage, networks, and other fundamental computing resources where the consumer is able to deploy and run arbitrary software, which can include operating systems and applications.

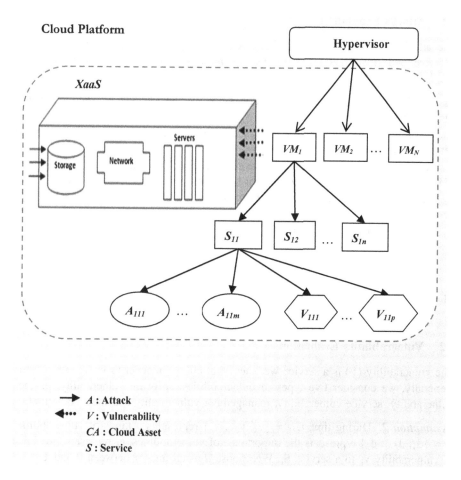

Fig. 4. A *CSQA* model for *XaaS*

In fact, cloud offers anything as a service therefore we adopt the *XaaS* definition when *X* represents the service to provide to cloud customer [17]. The first purpose of this work is to evaluate quantitatively any *XaaS*. The *XaaS* can be a service deployed in a *VM* such as the *SaaS* or it can be many distributed services in different *VMs* such as *PaaS* and *IaaS*. The Fig. 4 illustrates the hierarchical model to evaluate the *XaaS* security level.

4 CSQA: Quantitative Assessment

Our *CSQA* is based on the *NSSA* conceptual model (Fig. 2). In this section, we focus on the *CSQA* situation evaluation. We will evaluate an *XaaS* exposed to attacks and vulnerabilities and which can be protected using security mechanisms. Next, we detail the quantitative assessment methods step by step.

4.1 Attacks Evaluation

The attack (*A*) is any action that jeopardizes the security of cyberspace, including effective and invalid attacks. We use assumptions, which are similar to Chen's conclusion [4].

Assumption 1. During time *t*, a cloud service can be affected by such number of attacks. We consider that $S = \{S_1, S_2, .., S_n\}$ represents *n* kinds of services and $A = \{a_1, a_2, .., a_m\}$ represents *m* kinds of attacks affecting a service S_i when *i* is an integer between **1** and *n*.

Each attack has an importance so we consider $B = \{b1, b2, .., b_m\}$ as the importance of the attack. We choose *b* as an integer between 0 and 1.

The c_j is a number between 0 and 1 which represents the value of the attack a_j faced by a service S_i.

The attacks assessment is *R(t,A)*:

$$R(t, A) = \sum_{j=0}^{m} C_j 10^{b_j} \tag{1}$$

The bigger the value is, the more insecure the service is.

4.2 Vulnerabilities Evaluation

The vulnerability (*V*) is a service weakness that allows attacker to violate its security. Generally, we consider five types of vulnerabilities: software vulnerability, protocol vulnerability, service vulnerability, management vulnerability and false configuration.

Assumption 2. During time t, $v = \{v_1, v_2, .. v_p\}$ represents *p* kinds of vulnerabilities. $D = \{d_1, d_2, .., d_p\}$ represents the importance of vulnerabilities. e_k represents the effect of vulnerability v_k to a service S_i. We choose *d* as an integer between 0 and 1.

The vulnerability assessment is $R(t,V)$:

$$R(t, V) = \sum_{k=0}^{p} e_k 10^{d_k} \qquad (2)$$

The bigger the value is, the more vulnerable the service is.

4.3   Security Services Evaluation

We consider the five security attributes of confidentiality, availability, integrity, authentication, non-repudiation defined by *DoD* [18] additionally to their security mechanisms.

Assumption 3. During time t, N = $\{N_1, N_2, ..., N_5\}$ represents five security attributes. M = $\{M_1, M_2, ..., M_q\}$ represents q kinds of security mechanisms. The β = $\{\beta_1, \beta_2, ..., \beta_5\}$ represents the weight of the security attributes. We choose β as an integer between 0 and 1.

g_i represents the effect of the security mechanism M_s (where s is an integer between 1 and q) to security attribute N_l for the service S_i. We choose g as an integer between 0 and 1.

The security services assessment is $R(t,M)$:

$$R(t, M) = \sum_{l=1}^{5} \beta_l \sum_{s=1}^{q} g_s \qquad (3)$$

The bigger the value is, the safer the service is.

According to Assumption 1, 2 and 3, current service security assessment $R(t,A,V, M)$ is calculated in formula (4):

$$R(t, A, V, M) = G \cdot \frac{R(t, A) R(t, V)}{R(t, M)} \qquad (4)$$

G is a normalization factor. This factor is introduced by the cloud security expert to reduce all the results if they have big values.

4.4   The *XaaS* Security Quantitative Assessment

To evaluate the *XaaS*, we consider all the *XaaS* levels in the *CSQA* model. So, we define:

- The W_S = $\{\alpha_1, \alpha_2, ..., \alpha_n\}$ representing the weights of all the cloud services.
- The W_{CA} = $\{\lambda_1, \lambda_2, ..., \lambda_N\}$ representing the weights of the cloud assets.

These weights are defined by the cloud security expert.
Now, to calculate the *XaaS* security index, we have:

- Based on (1), the *XaaS* attacks assessment $R_{XaaS}(t,A)$ is:

$$R_{XaaS}(t, A) = \sum_{h=1}^{N} \lambda_h \sum_{i=1}^{n} \alpha_i \sum_{j=0}^{m} c_j 10^{b_j} \tag{5}$$

- Based on (2), the *XaaS* vulnerabilities assessment $R_{XaaS}(t,V)$ is:

$$R_{XaaS}(t, V) = \sum_{h=1}^{N} \lambda_h \sum_{i=1}^{n} \alpha_i \sum_{k=0}^{p} e_k 10^{d_k} \tag{6}$$

- Based on (3), *XaaS* security mechanism assessment $R_{XaaS}(t,M)$ is:

$$R_{XaaS}(t, V) = R(t, M) \tag{7}$$

As a conclusion based on (5), (6) and (7), we calculate the $R_{XaaS}(t,A,V,M)$ as follows:

$$R_{XaaS}(t, A, V, M) = G. \frac{R_{XaaS}(t, A) \, R_{XaaS}(t, V)}{R_{XaaS}(t, M)} \tag{8}$$

5 *CSQA* Framework Based on *MA/WS* Interaction

Based on both *NSSA* conceptual model (Fig. 2) and the *CSQA* model (Fig. 4), we suggest a *CSQA* framework (Fig. 5) to calculate the *XaaS* cloud security.

This framework is based on the mobile agents platform [19]. It is a distributed platform that offers a mobile code (corresponding to the *MA*) able to migrate from one host to another in order to collect data and to distribute computing charge.

We detail below the framework components and functions.

5.1 CSQA Framework Components

The CSQA is composed of 4 principal components:

- Cloud Asset (*CA*): it represents a cloud resource such as *VMs*, hypervisor, etc.
- Cloud Sensors (*CS*): It is a device or software deployed in a *CA* to detect attacks and vulnerabilities. The *CSQA* framework deploys two types of sensor:
  - *VM Sensors:* sensors can be a log manager for system, for software, for service, a *HIDS/HIPS*, anti-virus, etc.
  - *Hypervisor Sensors*: It can be a log manager sensor used to collect all VMs behavior. It can be also a *NIDS/NIPS* to detect and prevent all cloud attacks.

  The *CS* store their outputs in a database. These *CS* outputs are used to calculate the security situation index.

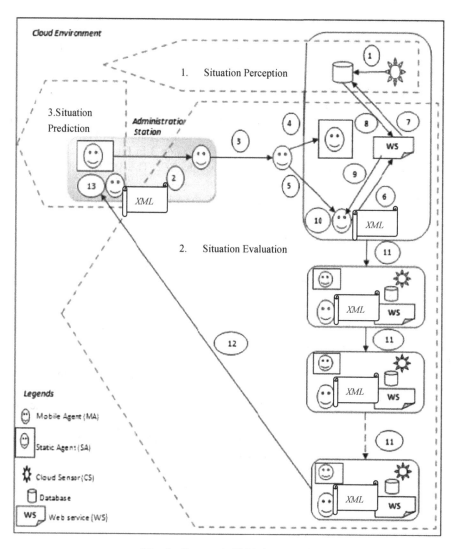

Fig. 5. Proposed *CSQA* framework

- Web service (*WS*) [20]: The *WS* is deployed in a *CA*. It is a middleware between *CS* databases and mobile agents (*MA*). It gives the required information to the *MA*.
- Mobile Agents Platform [19]: It is a distributed platform composed of two types of agents:
  - Static Agent (*SA*) deployed in the *CA* to dispatch and to receive the *MA*. It requires authentication step to accept *MA*.
  - Mobile Agent (*MA*) migrates from the administration station to the *CA* in order to calculate the *XaaS* security index.

5.2   CSQA Framework Functions

The *CSQA* is a three-phases flow as presented in the *NSSA* conceptual model:

- *Phase 1: Situation Perception*
    - Step 1: The *CS* detect service attacks and vulnerabilities and store them in their databases (step 1 in Fig. 5).
    - Step 2: The *SA* deployed in the administration station sends a *MA* to the different Cloud services to evaluate cloud security situation (step 2 in Fig. 5).
    - Step 3: The *MA* migrates to the first service (step 3 in Fig. 5).
    - Step 4: a *SA* deployed in the *CA* receives the *MA*. The *MA* should be authenticated (step 4 in Fig. 5).
    - Step 5: If the authentication is successful, the *SA* accepts the *MA* to continue its mission (step 5 in Fig. 5).
    - Step 6: After being accepted by the *SA*, The *MA* invokes the *WS* to collect services attacks and vulnerabilities (step 6 in Fig. 5).
    - Step 7: The *WS* requests the *CS* databases to aggregate all the information required by the *MA* (step 7 in Fig. 5).
    - Step 8: The databases responds the *WS* within service attacks and vulnerabilities (step 8 in Fig. 5).
    - Step 9: The *WS* responds the *MA* by the required information (step 9 in Fig. 5). The *MA* stores data in an *XML* file.
- *Phase 2: Situation Evaluation*
    - Step 10: The *MA* calculates the first service security index corresponding to the *XaaS* to evaluate.
    - Step 11: After evaluating all the *XaaS* relative services in the first *CA*, the *MA* migrates to other *CA* and repeats the same steps (step 2, 3, 4, 5, 6, 7, 8, 9).
    - Step 12: After compleeting its mission, the *MA* returns back to the administration station with the $R_{XaaS}(t, A, V, M)$ (step 12 in Fig. 5).
- *Phase 3: Situation Prediction*
    - Step 13: Based on the historical situation values, the *SA* plots situation map and forecasts the future situation (step 13 in Fig. 4).

6   Case Study

In a previous work, we have proposed a secure cloud architecture design [21] that we deployed in this work on proxmox hypervisor. The architecture is a small *IaaS* composed of two ubuntu *VMs* (Fig. 6). It is a proof of concept for our *CSQA* model.

In our *IaaS* framework:

- We deploy a snort [22] as an *IDS/IPS* in the hypervisor and in the *VMs*. Then, we implement a *WS* in each *VM* to interact with the snort database.
- In the first *VM*, we deploy an apache server. We simulate a *DDOS* attack to break down the apache server.
- In the second *VM*, we open a telnet port to be detected as a vulnerability.

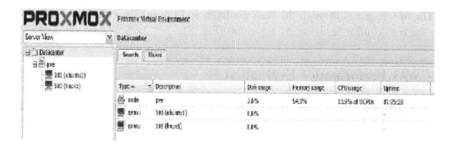

Fig. 6. *IaaS* based on Proxmox

- We used the *Aglet* platform [23] to implement the *MA* approach. We deploy the Tahiti which is the aglet server in the administration station. Then, we activate a *SA* in all the *VM*s. Finally, we send The *MA* to calculate the *IaaS* security index (Fig. 7).

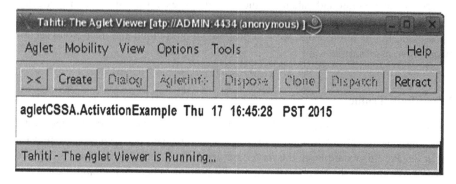

Fig. 7. *MA* migration from the Tahiti server

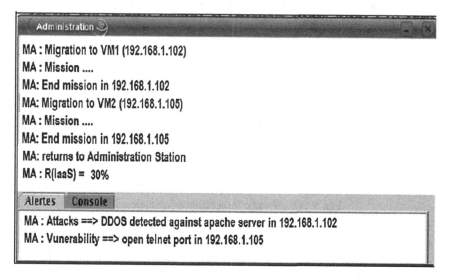

Fig. 8. *MA* mission in the *IaaS*

The *MA* migrates through the IaaS platform and returns back to the administrator with the results after applying its algorithm (Fig. 8).

7   *CSQA* Performance Evaluation

The *CSQA* framework evaluation is based on 4 criterions: the scalability of the *CSQA* model, the cloud traffic, the secure communication and the autonomous and asynchronous framework.

7.1   The *CSQA* Scalability

In this work, the *CSQA* proves its capability to quantitatively evaluate the security level of any *XaaS* in the cloud. It can be extended to cover all cloud assets (*VMs* and hypervisor) evaluation and calculates the security index for an intra-cloud platform.

In this case, the hypervisor security index will be a part of the intra-cloud security assessment. In our future work, we will propose an intra-cloud and inter-cloud security quantitative assessment.

7.2   Cloud Traffic

Our *CSQA* aims to decrease the cloud traffic. It benefits from the distributed system advantages based on *MA* to reduce the requests exchanged between the different *CA*.

Let's take this example: in the *IaaS* case, we have N services each one is deployed in a *VM*. We studied the number of requests for both the Client/Server approach and the mobile agent approach.

- Client/Server approach (Fig. 9): To evaluate the $R_{XaaS}(t,A,V,M)$, we noted that there are $2N$ requests to exchange.

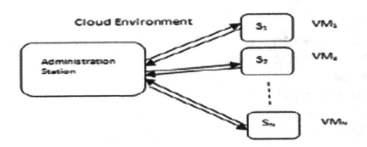

Fig. 9. Cloud traffic for a client/server approach

- Mobile agent approach (Fig. 10): the *MA* migrates from a *VM* to another in order to evaluate the R_{IaaS}. It returns back to the administration station with final assessment. So there are $N + 1$ exchanged requests in the cloud.

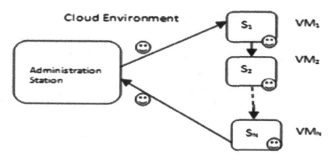

Fig. 10. Cloud traffic for a *MA* approach

The *MA* used in the *CSQA* framework reduces significantly the cloud traffic compared to the client/server model. When the service number increases the gain will be more important (Fig. 11).

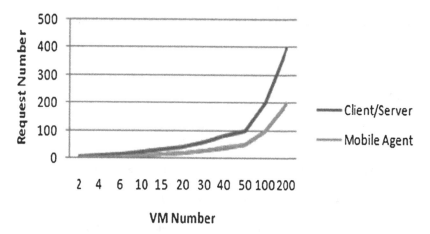

Fig. 11. Cloud traffic: *MA* approach versus client/server approach

7.3 Secure Communication

Our *CSQA* ensures a controlled and a secure access to cloud resources by two solutions:

- The authentication between agents is a mandatory step. Thus, the *MA* should exchange a password with the *SA* before being deployed in the cloud asset. If the authentication fails, the *MA* will be rejected.
- The *MA* can't have a direct access to cloud resources. It invokes *WS* to get necessary information to calculate the security index. In fact, the *WS* is considered as a middleware between *MA* and internal cloud resources such as databases, files, etc. So *MA/WS* interaction is full of advantages as studied in [24]. We benefit from these advantages to design our *CSQA* framework.

7.4   Autonomous and Asynchronous System

The autonomy is one of the most important properties of our *CSQA* framework due to the use of mobile agent platform: Independently from the administration station, the *MA* migrates from *CA* to another to accomplish its mission.

The *MA* platform is also asynchronous: if the *MA* sends the result to a non operational administration station, the results can be received when coming back operational.

Those two properties ensure an autonomous and asynchronous *CSQA* framework.

8   Conclusion

In this study, we firstly succeeded to suggest a *CSQA* model to calculate an *XaaS* security index in the cloud. Then, we explored the advantages of both, mobile agent approach and the *MA/WS* interaction to design a *CSQA* framework. This framework proves its capability to reduce cloud traffic, to secure communication, to be autonomous, asynchronous and scalable.

This work focus mainly on the situation evaluation step. In the future work, we will detail situation prediction step of the *CSQA* model.

References

1. Cloud Adoption Practices & Priorities Survey Report. https://downloads.cloudsecurity alliance.org/initiatives/surveys/capp/Cloud_Adoption_Practices_Priorities_Survey_Final.pdf. Accessed 16 Feb 2016
2. Endsley, M.R.: Design and evaluation for situation awareness enhancement. In: Human Factors Society, 32nd Annual Meeting, Santa Monica, CA (1988)
3. Bass, T., et al.: A glimpse into the future of ID. http://www.usenix.org/publications/login/1999-9/features/future.html
4. Chen, X.Z., et al.: Quantitative hierarchical threat evaluation model for network security. J. Softw. **17**(4), 885–897 (2006)
5. Jibao, L., et al.: Study of network security situation awareness model based on simple additive weight and grey theory. IEEE (2006)
6. Yong, Z., Xiaobin, T., Hongsheng, X.: A novel approach to network security situation awareness based on multiperspective analysis. In: 2007 International Conference on Computational Intelligence and Security, pp. 768–772. IEEE, December 2007
7. Xiaorong, C., Su, L., Mingxuan, L.: Research of network security situational assessment quantization based on mobile agent. Phys. Procedia **25**, 1701–1707 (2012)
8. Dastjerdi, A.V., Bakar, K.A., Tabatabaei, S.G.H.: Distributed intrusion detection in clouds using mobile agents. In: 3rd International Conference on Advanced Engineering Computing and Applications in Sciences (2009)
9. Toumi, H., Eddaoui, A., Talea, M.: Cooperative intrusion detection system framework using mobile agents for cloud computing. J. Theor. Appl. Inf. Technol. **70**(1) (2014)
10. Doelitzscher, F., et al.: An agent based business aware incident detection system for cloud environments. J. Cloud Comput. **1**(1) (2012)

11. Zargar, S.T., Takabi, H., Joshi, J.B.D.: DCDIDP: a distributed, collaborative, and data-driven intrusion detection and prevention framework for cloud computing environments. In: International Conference on Collaborative Computing: Networking, Applications and Worksharing (CollaborateCom), Orlando, Florida (2011)

12. Kamongi, P., et al.: Vulcan: vulnerability assessment framework for cloud computing. In: 7th International Conference on Software Security and Reliability (SERE). IEEE (2013)

13. Saripalli, P., Walters, B.: QUIRC: a quantitative impact and risk assessment framework for cloud security. In: 2010 IEEE 3rd International Conference on Cloud Computing (CLOUD). IEEE (2010)

14. Albakri, S.H., et al.: Security risk assessment framework for cloud computing environments. Secur. Commun. Netw. **7**(11), 2114–2124 (2014)

15. Sen, A., Madria, S.: Off-line risk assessment of cloud service provider. In: 2014 IEEE World Congress on Services (SERVICES). IEEE (2014)

16. Mell, P., Grance, T.: The NIST definition of cloud computing. NIST Special Publication 800–145 (Draft) (2011). Accessed 11 Oct 2015

17. Schaffer, H.E.: X as a service, cloud computing, and the need for good judgment. IT Prof. **11** (5), 4–5 (2009)

18. DoD Directive 3600.1, Information Operations, December 1996

19. Lange, D., Oshima, M.: Seven good reasons for mobile agents. Commun. ACM (1999)

20. Lemahieu, W.: Web service description, advertising and discovery: WSDL and beyond. In: Vandenbulcke, J., Snoeck, M. (eds.) Leuven University Press (2001)

21. Khaldi, A., Karoui, K., Tanabene, N., Ben Ghzala, H.: A secure cloud computing architecture design. In: 2nd IEEE International Conference on Mobile Cloud Computing, Services, and Engineering (MobileCloud), pp. 289–294. IEEE (2014)

22. Snort, January 2016. http://www.snort.org/

23. Aglet, January 2016. http://aglets.sourceforge.net/

24. Ben Ftima, F., Karoui, K.: Interaction mobile agents - web services. In: Encyclopedia of Multimedia Technology and Networking, 2 edn., pp. 717–725 (2007)

A Middleware to Allow Fine-Grained Access Control of Twitter Applications

Francesco Buccafurri$^{(\boxtimes)}$, Gianluca Lax,
Serena Nicolazzo, and Antonino Nocera

DIIES, University Mediterranea of Reggio Calabria,
Via Graziella, Località Feo di Vito, 89122 Reggio Calabria, Italy
{bucca,lax,s.nicolazzo,a.nocera}@unirc.it

Abstract. Mobile applications security is nowadays one of the most important topics in the field of information security, due to their pervasivity in the people's life. Among mobile applications, those that interact with social network profiles, have a great potential for development, as they intercept another powerful asset of the today cyberspace. However, one of the problems that can limit the diffusion of social network applications is the lack of fine-grained control when an application use the APIs of a social network to access a profile. For instance, in Twitter, the supported access control policy is basically on/off, so that if a (third party) application needs the right to write in a user profile, the user is enforced to grant this right with no restriction in the entire profile. This enables a large set of security threats and can make (even inexpert) users reluctant to run these applications. To overcome this problem, we propose an effective solution working for Android Twitter applications based on a middleware approach. The proposed solution enables other possible benefits, as anomaly-based malware detection leveraging API-call patterns, and it can be extended to a multiple social network scenario.

Keywords: Application security · Fine-grained access control · Android · Twitter · OAuth

1 Introduction

One of the most important topics in the field of information security is mobile applications security. Indeed, the diffusion of smartphones has rapidly modified people habits, by allowing people to be always connected via social networks, but also by enabling ubiquitous and pervasive applications. This has dramatically enlarged the attack surface making each user of mobile applications a potential victim of cyber-attacks. Users can be affected in several ways, including data theft or corruption, annoyance, device damage or location tracking [14,29].

Among mobile applications, social network applications have a great potential for development, because they use social networks as a platform for information sharing, user-centered content production, and interoperability [15,17,18,40]. Activities and interests can be shared among people, making applications aware of the social factor, which is often determinant for the success and the effectiveness of the application.

© Springer International Publishing AG 2016
S. Boumerdassi et al. (Eds.): MSPN 2016, LNCS 10026, pp. 168–182, 2016.
DOI: 10.1007/978-3-319-50463-6_14

However, one of the problems that can limit the full diffusion of social network applications is that, in the current real-life scenario, the most social network providers do not support a fine-grained control when an application use the APIs of the social network to access a profile. This is the case of Twitter, in which the supported access control policy is roughly on/off. For an instance, if an application needs the right to tweet within a given user profile, the user is enforced to grant to the application the right to modify (even to delete) any information included in his profile with no restriction. This enables a large set of security threats because users cannot fully control what their applications do in their Twitter profile, so they cannot block their potential malicious behavior.

We can argue that the effect of the above policy may strongly limit the growth and the diffusion of social network applications, despite their strategic utility. For instance, even for an inexpert user, it can appear inopportune to run an application asking for a complete control of his Twitter profile.

To overcome this problem, we propose in this paper an effective solution working for Android Twitter applications based on a middleware approach. This middleware is spread-out between the smartphone operating system and a server-side platform, and allows the definition of a fine-grained access control model to protect the end-user. The client-side middleware is thought as an application that *hijacks* the Twitter API calls to the server-side middleware, where calls are implemented according to the access control rules. However, a large-scale adoption of our solution could be implemented as an extension of Android itself. Interestingly, our solution enables other possible benefits, as anomaly-based malware detection leveraging API-call patterns. Moreover, by using the model proposed in [16], it can be extended to a multiple social network scenario.

The paper is organized as follows. We start by providing in Sect. 2 the background on OAuth, the authentication protocol used in the scenario considered. Next, in Sect. 3, we motivate the problem and show an overview of the characteristics that a solution should have. The description of our proposal and its implementation is given in Sect. 4. The comparison with related work is provided in Sect. 5. Finally, we conclude the paper by discussing some interesting issues and by drawing our conclusions.

2   Background

In this section, we provide the technical knowledge necessary to understand our proposal. It concerns the authentication mechanisms on social networks.

Most of online social networks have embraced the paradigm of Software as Service (SaaS) meaning that all features are available to end users as a service through the Internet. The services offered to users are delivered as a set of Application Programming Interfaces (APIs) available via the http/s protocol. However, one of the main issue is the necessity of providing authentication mechanisms that do not necessary require the direct log into the social network for accessing its data and functions through external services. For this purpose, the authentication of many social networks, such as Facebook, Twitter, Google+, is based on the OAuth 1.0a protocol [31] described in RFC 5849, which has been replaced by the new OAuth 2.0 authorization framework [32].

Basically, OAuth allows a third-party application to gain limited access to an http/s service without users expose their password. To do this, it provides a token-based mechanism so that an application can access data on in a social network profile if a valid token is provided. Such a token is automatically delivered to the application directly from the social network after the user manually approves it. In the latest version of the Twitter API (v1.1), two forms of authentication, both based on Oauth 1.0a, are allowed: *Application–Only* Authentication, where only the application is authenticated and acts without using a user context, and *Application–Users* Authentication, which allows the application to act on the user's behalf. The former is the basic authentication and allows it to call APIs and retrieve public information on the social network. Clearly, in this case, the application cannot obtain any private information of any user and cannot perform actions, such as tweeting, on the user's behalf on Twitter. Now, we focus our attention and technical discussion on the case of Twitter. From a technical point of view, Twitter authentication scheme is based on the use of two Twitter API keys, namely the *consumer key* and the *consumer secret*. Specifically, the consumer key is used to identify the application itself, whereas the consumer secret, in combination with the consumer key, allows the application to perform authenticated requests on its behalf.

The detailed flow is as follows: First, the application encodes its consumer key and secret and performs an https POST request to the link https://api.twitter.com/oauth2/token to send these credentials to the `oauth2/token` endpoint. Then, the `oauth2/token` validates them and replies with a *bearer token*. Finally, the application uses the bearer token for the future interactions with the Twitter APIs. Clearly, this authentication scheme does not support services or data requiring a user context. The Application–Users Authentication is the most common authentication scheme and works by authenticating both the application (or external service) and the user. The mechanism allows users to explicitly declare their willing to provide the application with a token for opening a context on their behalf. This step is allowed only if the application is registered on Twitter and, hence, is equipped with a valid consumer key. The basic requirement for this authentication scheme is an *access token*, which allows the application to operate on behalf of the user whom this access token belongs to. There are mainly three authentication mechanisms to obtain the user token and are related to the specific functionality the application wants to provide. Specifically, the possible authentication mechanisms are: *Sign in with Twitter*, *3-legged OAuth* and *PIN-based OAuth*. As for the first mechanism, in this case the application uses Twitter account information to identify a user, thus allowing a fast log into its services. The application performs a POST requests, using its consumer information as credentials, to the oauth/request_token endpoint via the link https://api.twitter.com/oauth/request_token. This request must embed as parameter a callback URL where the user is redirected after the authentication on Twitter. In the response, a *request token* is returned, which is used as a GET parameter to redirect the user (via browser) to the link https://api.twitter.com/oauth/authenticate. This way, the user can authenticate on Twitter, which, if the login procedure succeeds, sends a response containing a *verify token* univocally associated with the request token to the callback URL. Finally, to obtain a valid *access token*, allowing the retrieval of the account information for the user via the `account/verify_credentials` endpoint, the application performs a POST request using the verify token as parameter to link https://api.twitter.com/oauth/access_token. The `account/verify_credentials` endpoint can be accessed through the link https://api.twitter.com/1.1/account/verify_credentials.json.

The second authentication mechanism *3-legged OAuth* is used when an application wants the privileges to act on the user's behalf. To do this, the user is redirected on Twitter to authorize the application. This mechanism is almost identical to the previous one. The only significant difference is that the access token obtained with this procedure can only be used once. In the case the application cannot directly make use of a browser to obtain the access token, it can use the *PIN-based OAuth*. Also this mechanism is similar to the first. However, since the application cannot access a Web browser, a URL is returned by Twitter in the last step of the procedure. This is shown to the user who can now use a browser to access this link and authenticate on Twitter. If the authentication succeeds, than a Web page with a 7-digit code is shown to the user. Now, the user has to input this code to the application (that should be waiting for this input) and the application can complete the procedure to obtain the access token by sending a final request to the Twitter endpoint `oauth/access_token` with the 7-digit code as parameter.

Once the application has been authenticated, Twitter can grant it one of the following privileges (chosen before the token generation): *(1) Read only*, meaning that the application can only read information on the user timeline and account *(2) Read & Write*, meaning that the application can perform all reading activities along with writing/deleting contents and profile statuses as well as sending direct messages to other users. Clearly, privilege 2 subsumes privilege 1.

3   Problem Formulation and Desiderata

In this section, we motivate the problem by real-life examples and sketch the characteristics that a solution of this problem should have. To better illustrate how our approach works, we make explicit reference to a scenario in which a mobile application requires the access to Twitter information via services provided by the social network (according to the SaaS paradigm). This scenario is very common, as there are a lot of third-party mobile applications that offer enhanced functionalities extending Twitter basic services.

3.1   Motivations

The first example we present concerns the case of an application allowing users to write tweets with more than 140 characters. There are several ways to achieve this goal: by automatically creating a short URL linking the whole user tweet, or by posting the long message as an image, or by splitting the original tweet into a series of smaller linked tweets. In this case, an application offering this feature should act on behalf of the user and, therefore, should have the *Read & Write* privilege.

A more complex and interesting example of the SaaS paradigm application is given by the use of the social signature [36]. Social signature is a recent proposal implementing electronic signature by means of Twitter. This social network is used as "device" enabling the generation of the signature and also as a trusted-third-party allowing signature sharing. Signers use their mobile phone to generate the signature and to post a suitable tweet on their own account. The tweet shows the digest of the signed document along with other information necessary to guarantee authenticity and non-repudiation. In this case, the data to be posted on Twitter cannot be manually generated by the user. Again, the mobile application implementing the social signature protocol should be authorized by the user to run Twitter services with `Read & Write` privilege. The most relevant side-effect of this authorization is that, once an application obtains the consumer's access token, it may perform almost everything on the user profile, although it needs only a subset of those permissions. The examples described above clearly show situations in which the low granularity of Twitter access control model gives rise to possible security threats because the least privilege principle is not accomplished.

3.2   Solution Requirements

To address the problem described above, a possible solution should satisfy the following requirements:

Fine-Grained Privileges on Twitter Services. The solution should allow the user to explicitly define the specific functionalities allowed to a mobile application on his Twitter account. Specifically, the user should be able to directly specify the set of API methods that the application is allowed to use. In the first example of Sect. 3, the user may allow the application to only use the `POST statuses/update` method instead of granting any call to other not required writing methods, such as `POST statuses/destroy/:id`, which allows the deletion of a previous tweet or `POST account/update_profile` allowing the update of personal profile information such as the user contact name.

Device-Independent Policy Definition. Once the user defines a content access policy to the Twitter functionalities for a given service, this should be applied to any application on each specific device that the user wants to bound with these policies. This aspect has a heavy impact on the solution architecture because access control rules must be available outside the user device.

Anomaly-Based Detection of Malicious Usage of the Service. Although the user is able to specify fine-grained privileges, the misuse access privileges by an application is still possible. A valid solution should be able to detect this illegal behavior by analyzing how an application makes use of Twitter API's methods.

We will see in the following how to build a solution able to ensure such requirements.

4   System Architecture

In this section, we describe our solution and the general architecture of its implementation. Specifically, to satisfy all requirements described in Sect. 3.2, we identify two main components for our system: *(i)* An Android service that monitors all traffic toward Twitter from the user device; *(ii)* a remote middleware that communicates with the Android service on the user device and verifies whether the activities performed by third-party applications match the user access control rules.

A schema of the proposed framework is shown in Fig. 1.

The basic idea is that every call to a method of the Twitter APIs that is generated by a third-party application running on the user device is intercepted and analyzed. To do so, we implement an application requiring root privileges on Android and monitoring all https traffic towards Twitter. When a communication towards Twitter is identified, the flow is redirected to our remote middleware, which will act as an intermediate by performing the requests to Twitter on behalf of the application that originated the communication. The middleware implements all the functionalities to satisfy the requirements of Sect. 3.2. In the next sections, we will describe the two components in more detail.

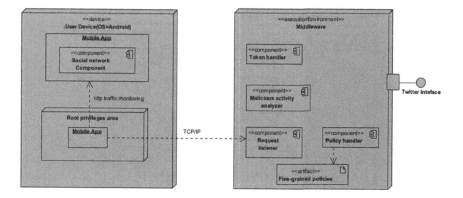

Fig. 1. Deployment diagram of the proposed framework.

4.1   The Android Service

To achieve the goal our system must take control over the communication towards Twitter originated by the user device. For this purpose, we provide the device with an application, implemented as an Android service, that has to monitor all the Web traffic. However, as stated in the Android specifications [2], the execution of an Android application follows the *principle of least privilege* [25] for which each application runs in a separated Linux process inside an independent virtual machine. Moreover, each Linux process, associated with an Android application, is executed by a different Linux user, who can access only the information in its isolated address space. The mechanism underlying this specification ensures that no direct interaction may occur between different applications. Figure 2 schematizes the concept outlined above.

To bypass this *sandbox* security strategy of Android, we have to give to our application the root privileges. This way, it can access all the information available and makes use of the underlying Linux operative system functionalities. The target of our root-level application is very similar to that of some other famous applications, such as [3,11,13]. Observe that, the practice of rooting an Android phone is widely accepted especially for security solutions, such as [4,7,10]. Moreover, it is also common to run applications that require special privileges to bypass carrier installed software, overclock and underclock the

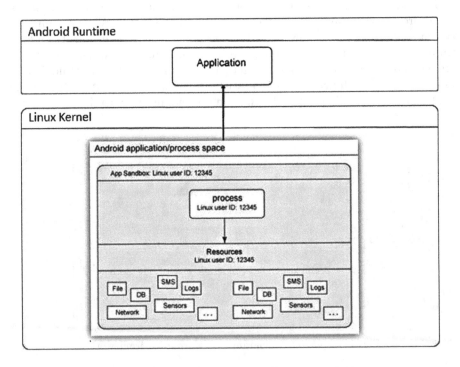

Fig. 2. Android application execution representation.

processor [12], modify system files [5], or even more trivial actions like system setting customization [8]. However, the main difference is that our application has not only to intercept any communication towards Twitter starting from the device, but also has to redirect the communication towards our middleware server. Our middleware will perform all the https requests to Twitter on behalf of the user application and will return responses back to the application on behalf of Twitter. To achieve this goal our application does the following: *(i)* jeopardizes the PKI trust chain by installing a self-signed fake root certificate (this is possible with root privilege from Android 4.0 on [1]); *(ii)* makes use of the powerful Linux firewall underlying the Android system, namely `iptables` [9]. By properly setting up rules on `iptables`, it is possible to redirect any *TCP* traffic accordingly to our strategy. Observe that, even though it could be in principle considered a break of security, the jeopardization of the PKI trust chain is, in fact, a practical solution. Indeed, it is commonly adopted in supervised contexts (such as, enterprise environments), in which a proxy servers perform traffic flow analysis and monitoring [6,26]. To do so, we started from the source code of *DroidWall - Android Firewall* that is an open source project for a front-end application to the Linux firewall [4]. The aim of DroidWall is to restrict the set of applications that are allowed to access the device data networks, therefore we suitable modify its code to implement our solution.

4.2   The Middleware Service

This component of our system is in charge of managing the authorization policies, defined by the user, to the Twitter API's contents. Specifically, all the traffic generated by any third-party applications on any user-device is redirected, by our Android service, to this external component. In this component we have the following modules:

Request Listener. This module receives all the https requests originated by our Android service running on any device.

Token Handler. The aim of this module is to handle the interaction with the OAuth authentication protocol. In the initialization step, our application requires the access to Twitter services by issuing an *Application–Users Authentication* via OAuth protocol (see Sect. 2 for further details). The output of this procedure is that the user is prompted to authorize the application to access his Twitter account; in this phase, the application requires also the authorization type (which is either *Read only* or *Read & Write*). At the end of this procedure, in the case of positive authentication, the application is provided with a (long-term) token that can be used to speed up the authentication procedure in the next communications with Twitter. The basic strategy of our approach is to replace the original request of Application–Users Authentication performed by the third-party application on the user device with a corresponding request generated by the middleware itself. This way, the middleware will obtain the access token, hereafter referred as *master* token, from Twitter and will perform all the next requests on behalf of the third-party application. On the other side, this component generates a *sub*-token, associated with the original *master* token, that will be sent back to the application. As specified by the OAuth protocol, the application will use this *sub*-token in the call to any Twitter API's method.

Policy Handler. This module has two main functions: *(i)* to store a set of access control rules for each application and *(ii)* to provide the right mapping between any sub-token and the corresponding rules. The implementation of fine-grained policies is orthogonal to the approach described in this paper, therefore it is possible to follow any existing approach, such as [28,39], even though a specific model would be desirable (this is matter of our future work). A possible example of an access policy could be the following. Consider the case in which a user wants to prevent a third application from reading his private direct messages. In this case, the Twitter APIs handling direct messaging are the following:

- GET direct_messages/sent. The call to this method returns the 20 latest direct messages sent by the authenticating user together with information about the sender and the recipient user.
- GET direct_messages/show. This method returns the direct message whose identifier is specified in the *id* parameter, along with information about the objects, the sender and the recipient.
- POST direct_messages/destroy. Through this API method, the application can destroy a direct message received by the authenticating user, by specifying the message identifier in the required *id* parameter.
- POST direct_messages/new. This API method allows the application to send a new direct message to a specific user on behalf of the authenticating user.

Clearly, the user could not want to manually specify all the API methods that has to be blocked to achieve his goal. In this case, he can give only the general command ⟨*direct_message, deny*⟩. This module will translate this command into an XML-serialized rule, which will propagate the deny command to all the sub-elements (i.e., the API methods referring to the direct messaging feature) of the element DirectMessage. Clearly, because the user did not specify whether the rule has to affect reading or writing feature, it will be applied on both of them (see Listing 1.1).

Anomaly Detector. This module implements a further security mechanism allowing the detection of malicious behavior of third-party applications. By following an anomaly-based approach, the above task is accomplished on known applications for which a behavior fingerprint is stored in the middleware. To detect anomalies this module can take advantage of the approaches proposed in [37].

Twitter Interface. It is in charge of handling the communication between the middleware and Twitter. In particular, it forges the new messages to Twitter according to user policies and receives the query results from Twitter.

Listing 1.1. An example of a policy denying direct messaging.

```
[...]
<Policy>
<Read allow="False">
  <DirectMessage allow="False">
    <sent allow="False">
      <API>GET direct\_messages\/sent</API>
    </sent>
    <show allow="False">
      <API>GET direct\_messages\/show</API>
    </show>
  </DirectMessages>
</Read>
<Write allow="False">
  <DirectMessages allow="False">
    <Destroy allow="False">
      <API>POST direct\_messages\/destroy</API>
    </Destroy>
    <PostNew allow="False">
      <API>POST direct\_messages\/new</API>
    </PostNew>
  </DirectMessages>
</Write>
</Policy>
[...]
```

4.3 Protocol

In this section we provide some further details about the protocol adopted by the entities involved in our system to fulfill our solution tasks.

As stated in Sect. 2, OAuth protocol works by performing an initial strong authentication which allows the association of the user and the application that will performs queries on his behalf. The output of this procedure is that the application will be provided with an access *master* token through which it can interact with protected Twitter-user resources on his behalf. Our protocol interposes, between the application that wants to use external APIs and Twitter, both an Android service and a middleware. Hence, standing in the way of all user application requests, our middleware is able to analyze and send them to Twitter only if they match user access policies. Figure 3 describes our protocol among Twitter, the middleware and the Android device.

In the first phase, the Android service monitors the original request of Twitter APIs performed by the third-party application on the user device (1). Observe that, all the calls to methods of the twitter APIs are performed via the https protocol. Moreover, they share the same structure, i.e.: `https://api.twitter.com/[version]/[API method]/[parameters]`. Therefore, once our Android service (see Sect. 4) monitors all https traffic and redirect to the middleware components all traffic originally intended for Twitter (2). The listener module of the middleware receives all requests and activates the `Token Handler` and the `Policy Handler` which process this request and forward it to the `Twitter Interface` module if it matches the user policies (3). Also in this step, the `Anomaly Detector` module is activated and analyzes the request by

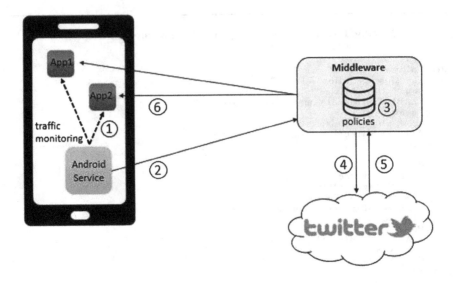

Fig. 3. A communication diagram of the entities involved in our system.

combining it with information on previous activities from the same third-party application to identify possible malicious behaviors. After that, in the case of positive matching, this module builds a similar request and sends it to Twitter in order to access the needed user resources (4). In a second phase, data flow back by following the same path described above. Observe that, in this way, the `Policy Handler` module could perform further filters on the obtained data to match user preferences in the case the used Twitter API method would not allow the right grain level (5) (for instance, read only the tweets of a given profile). We remark that this feature could not be supported by following a simple firewall-based approach. Finally, the filtered data are sent back to the Android device (6).

5   Related Work

Nowadays the lack of security in mobile devices is a critical issue because of the pervasive diffusion of these systems [21,33,35,38].

To solve this problem, several researchers investigated on fine-grained policy enforcement [19,23,41,42]. In [23], the authors define a system called CRePE able to enforce fine-grained policies depending on the context of the smartphone. According to the combination of the value of some variables (e.g. location, time, temperature, noise, the presence of other devices) that can define a particular context, the mobile device changes its settings and, for instance, it runs only a restricted set of applications or it switches off/on some interfaces. In [41], the authors create a framework to provide install-time permission-granting policies and runtime inter-application communication policies. This allows a specific application to change its runtime behavior, specifically: which external interface to use, which applications can access its interfaces, and how they can use its interfaces. Always in the field of fine-grained authorization, an interesting work is presented in [42]. In this paper, the authors introduce DAuth an authorization protocol that allows fine-grained control of access permissions in the context of application delegation. In particular, if an application is composed of more than one external component (i.e., different weblets), then a potential risk can occur. The countermeasure proposed in this work, is the definition of a new SDK to allow the design of applications where the external components have different policies w.r.t. the on-device ones. The authors of [19] try to unify distributed component models under a common meta-model for the purpose of creating a platform independent model for the access control patterns. Hence, the access control to application resources by the different components is handled through an orthogonal security aware middleware, thus allowing a plug and play component environment. All the approaches above assume that the applications that want to leverage their frameworks must be complaint to their protocols. This requirement is not needed by our solution that can be applied to any existing third-party application.

The problem of defining fine-grain authorization to access resources is a critical issue in all the environments in which remote sharing of application services is needed. For instance, [34] deals with the issue of guaranteeing fine-grain authorization for resource management in virtual organizations within grid environments. In particular, the authors of this paper extend the Globus Toolkit's (GT2) resource management mechanism [24] enriching community-wide policies with resource-specific policies (determined by the resource owner) and allowing a fine-grain control of how resources are used. Also the authors of [20] propose a solution for fine-grain access control in dynamic large scale grid environments focusing on security aspects.

As for mobile environment, the ability to install third-party applications poses serious security concerns that raise the attention of lots of researchers. In particular, the authors of [39] present a policy enforcement framework for Android called Apex. This framework allows users to selectively make decisions about permissions on their device rather than automating the decisions based on the policies of remote owners and allows finer-granular control over usage. In [27], the Kirin security service for Android is proposed. This system certifies an application according with a set of security rules, thus performing lightweight certification of applications and mitigating malware at install time. A relevant tool to perform automated security certification checking of Android applications as they are installed is SCANDROID [30]. It analyses whether data flows through an application are consistent with specifications extracted from the corresponding manifests.

Finally, since our model modifies the regular OAuth protocol flow in order to guarantee more granular choices on what an application can do with APIs it is useful to notice that, in this context, some extensions to the OAuth 2.0 authorization enabling an enhanced use of permissions are presented in [22,43].

Specifically, the focus of [43] is to provide a mechanism that computes permission ratings based on a recommendation model leveraging previous user decisions and application requests in order to enhance user privacy.

6 Discussion and Conclusion

In this paper, we have presented a concrete solution working for Android social applications using Twitter to support a fine-grained access control to the Twitter user profile. Besides the original approach we have proposed in this work, we have solved all the technical issues arising from its implementation. We have chosen Android and Twitter as they represents a large portion of the mobile/social market. However, the solution could be extended to other mobile platforms and social networks. To do this, we could exploit some previous results supporting multiple-social-network software development [16]. This is subject of our future work. Another direction towards we plan to investigate is the definition of a specific access control model, which, in this paper, we have considered as an orthogonal problem.

Acknowledgment. This work has been partially supported by the Program "Programma Operativo Nazionale Ricerca e Competitività" 2007–2013, Distretto Tecnologico CyberSecurity funded by the Italian Ministry of Education, University and Research.

References

1. Security SSL. http://developer.android.com/training/articles/security-ssl.html# Concepts
2. Android Developers (2015). https://developer.android.com/index.html
3. bitShark (2016). https://play.google.com/store/apps/details?id=blake.hamilton. bitshark
4. DroidWall (2016). https://code.google.com/p/droidwall/
5. Dumpster image and video restore (2016). https://play.google.com/store/apps/ details?id=com.baloota.dumpster
6. Firewall analyzer (2016). https://www.manageengine.com/products/firewall/ employee-internet-monitoring.html
7. Firewall pk+ (2016). https://play.google.com/store/apps/details?id=com. ikramshah.firewallpk
8. Gravitybox unlocker (2016). https://play.google.com/store/apps/details?id=com. ceco.gravitybox.unlocker
9. iptables (2016). http://www.netfilter.org/projects/iptables/
10. Mobile security and antivirus (2016). https://play.google.com/store/apps/details? id=com.avast.android.mobilesecurity
11. Network Log (2016). https://play.google.com/store/apps/details?id=com. googlecode.networklog
12. Setcpu for root users (2016). https://play.google.com/store/apps/details?id=com. mhuang.overclocking

13. SniffDroid (2016). https://play.google.com/store/apps/details?id=com.serious. sniffdroid

14. Buccafurri, F., Lax, G., Nicolazzo, S., Nocera, A.: A privacy-preserving solution for tracking people in critical environments. In: Proceedings of International Workshop on Computers, Software & Applications (COMPSAC 2014), pp. 146–151. IEEE Computer Society, Västerås (2014)

15. Buccafurri, F., Lax, G., Nicolazzo, S., Nocera, A.: Comparing Twitter and Facebook user behavior: privacy and other aspects. Comput. Hum. Behav. **52**, 87–95 (2015)

16. Buccafurri, F., Lax, G., Nicolazzo, S., Nocera, A.: A model to support design and development of multiple-social-network applications. Inf. Sci. **331**, 99–119 (2016)

17. Buccafurri, F., Lax, G., Nicolazzo, S., Nocera, A., Ursino, D.: Measuring betweenness centrality in social internetworking scenarios. In: Demey, Y.T., Panetto, H. (eds.) OTM 2013. LNCS, vol. 8186, pp. 666–673. Springer, Heidelberg (2013). doi:10.1007/978-3-642-41033-8_84

18. Buccafurri, F., Lax, G., Nicolazzo, S., Nocera, A., Ursino, D.: Driving global team formation in social networks to obtain diversity. In: Casteleyn, S., Rossi, G., Winckler, M. (eds.) ICWE 2014. LNCS, vol. 8541, pp. 410–419. Springer, Heidelberg (2014). doi:10.1007/978-3-319-08245-5_26

19. Burt, C.C., Bryant, B.R., Raje, R.R., Olson, A., Auguston, M.: Model driven security: unification of authorization models for fine-grain access control. In: Proceedings of 7th IEEE International Enterprise Distributed Object Computing Conference, pp. 159–171. IEEE (2003)

20. Butt, A.R., Adabala, S., Kapadia, N.H., Figueiredo, R., Fortes, J., et al.: Fine-grain access control for securing shared resources in computational grids. In: Proceedings of IEEE-IEE Vehicle Navigation and Information Systems Conference, 8-p. IEEE (1993)

21. Caviglione, L., Lalande, J.-F., Mazurczyk, W., Wendzel, S.: Analysis of human awareness of security, privacy threats in smart environments (2015). arXiv preprint arXiv:1502.00868

22. Cirani, S., Picone, M., Gonizzi, P., Veltri, L., Ferrari, G.: IoT-OAS: an OAuth-based authorization service architecture for secure services in IoT scenarios. IEEE Sens. J. **15**(2), 1224–1234 (2015)

23. Conti, M., Nguyen, V.T.N., Crispo, B.: CRePE: context-related policy enforcement for android. In: Burmester, M., Tsudik, G., Magliveras, S., Ilić, I. (eds.) ISC 2010. LNCS, vol. 6531, pp. 331–345. Springer, Heidelberg (2011). doi:10.1007/978-3-642-18178-8_29

24. Czajkowski, K., Foster, I., Karonis, N., Kesselman, C., Martin, S., Smith, W., Tuecke, S.: A resource management architecture for metacomputing systems. In: Feitelson, D.G., Rudolph, L. (eds.) JSSPP 1998. LNCS, vol. 1459, pp. 62–82. Springer, Heidelberg (1998). doi:10.1007/BFb0053981

25. Denning, P.J.: Fault tolerant operating systems. ACM Comput. Surv. (CSUR) **8**(4), 359–389 (1976)

26. Domingo-Pascual, J., Shavitt, Y., Uhlig, S.: Traffic Monitoring and Analysis, vol. 6613. Springer Science & Business Media, Heidelberg (2011)

27. Enck, W., Ongtang, M., McDaniel, P.: On lightweight mobile phone application certification. In: Proceedings of 16th ACM Conference on Computer and Communications Security, pp. 235–245. ACM (2009)

28. Ferrara, P., Tripp, O., Pistoia, M.: Morphdroid: fine-grained privacy verification. In: Proceedings of 31st Annual Computer Security Applications Conference, pp. 371–380. ACM (2015)

29. Ferreira, D., Kostakos, V., Beresford, A.R., Lindqvist, J., Dey, A.K.: Securacy: an empirical investigation of android applications network usage, privacy and security. In: Proceedings of 8th ACM Conference on Security and Privacy in Wireless and Mobile Networks (WiSec) (2015)
30. Fuchs, A.P., Chaudhuri, A., Foster, J.S.: Scandroid: automated security certification of android applications. Manuscript, University of Maryland, **2**(3), (2009). http://www.cs.umd.edu/avik/projects/scandroidascaa
31. Hammer-Lahav, E.: The OAuth 1.0 protocol (2010)
32. Hardt, D.: The OAuth 2.0 authorization framework (2012)
33. Jeon, W., Kim, J., Lee, Y., Won, D.: A practical analysis of smartphone security. In: Smith, M.J., Salvendy, G. (eds.) Human Interface 2011. LNCS, vol. 6771, pp. 311–320. Springer, Heidelberg (2011). doi:10.1007/978-3-642-21793-7_35
34. Keahey, K., Von, W.: Fine-grain authorization for resource management in the grid environment. In: Parashar, M. (ed.) GRID 2002. LNCS, vol. 2536, pp. 199–206. Springer, Heidelberg (2002). doi:10.1007/3-540-36133-2_18
35. La Polla, M., Martinelli, F., Sgandurra, D.: A survey on security for mobile devices. IEEE Commun. Surv. Tutor. **15**(1), 446–471 (2013)
36. Lax, G., Buccafurri, F., Nicolazzo, S., Nocera, A., Fotia, L.: A new approach for electronic signature. In: Proceedings of International Conference on Information Systems Security and Privacy (ICISSP 2016), Rome, IT (2016)
37. Maxion, R., Tan, K., et al.: Benchmarking anomaly-based detection systems. In: Proceedings of International Conference on Dependable Systems and Networks, DSN 2000, pp. 623–630. IEEE (2000)
38. Mylonas, A., Kastania, A., Gritzalis, D.: Delegate the smartphone user? Security awareness in smartphone platforms. Comput. Secur. **34**, 47–66 (2013)
39. Nauman, M., Khan, S., Zhang, X.: Apex: extending android permission model and enforcement with user-defined runtime constraints. In: Proceedings of 5th ACM Symposium on Information, Computer and Communications Security, pp. 328–332. ACM (2010)
40. Nikou, S., Bouwman, H.: Ubiquitous use of mobile social network services. Telematics Inform. **31**(3), 422–433 (2014)
41. Ongtang, M., McLaughlin, S., Enck, W., McDaniel, P.: Semantically rich application-centric security in android. Secur. Commun. Netw. **5**(6), 658–673 (2012)
42. Schiffman, J., Zhang, X., Gibbs, S.: Dauth: fine-grained authorization delegation for distributed web application consumers. In: IEEE International Symposium on Policies for Distributed Systems and Networks (POLICY), pp. 95–102. IEEE (2010)
43. Shehab, M., Marouf, S., Hudel, C.: RoAuth: recommendation based open authorization. In: Proceedings of 7th Symposium on Usable Privacy and Security, p. 11. ACM (2011)

AndroPatchApp: Taming Rogue Ads in Android

Vasilis Tsiakos and Constantinos Patsakis$^{(\boxtimes)}$

Department of Informatics, University of Piraeus, Piraeus, Greece
vasilis.tsiakos@gmail.com, kpatsak@unipi.qr

Abstract. Mobile applications have drastically changed the way that we use our mobile devices. The different sensors that are embedded allow novel user interaction and make them context-aware. However, the operating system of most mobile devices allows limited user configuration; the user does not have full access, in order to make them more secure. Despite this measure, the overall security and privacy of users cannot be considered adequate. While there are many tools for "rooted" devices, the choices for "out of the stock" devices are not that many, and more importantly, they are not that effective.

To address these shortcomings, AndroPatchApp takes a different approach. Instead of installing monitoring and detection apps in the operating system, AndroPatchApp embeds some security and privacy controls before installation, by generating a "sanitized" version of the app.

Keywords: Android · Mobile ads · Privacy

1 Introduction

The market of mobile devices has drastically changed the past few years with the adoption of Internet, the embedding of various sensors, but most importantly by allowing users to install mobile applications. The latter change led to many changes as many manufacturers had enabled Internet connectivity in their devices and shipped them with some sensors. However, users were restricted to using only a limited amount of applications which were usually pre-installed in the device. Allowing users to purchase applications and configure their devices according to their needs created a huge market shift.

While customization and application markets sound very interesting, allowing a user who is not very familiar with cyber security to arbitrarily change his device might expose him to great risks. To counter such exposure manufacturers ship their devices with an operating systems where users have limited access to its internals. Therefore, the customization that users can make and the applications that they can install is not arbitrary and significantly decreases the attack surface. Moreover, applications run in separate containers, each with different permissions, further decreasing the security and privacy issues to which a user is exposed to.

While the above may decrease the exposure, this does not by any chance mean that users are secure. On the contrary, users are exposed to many threats, and in many cases, due to manufacturers' update policies, they cannot perform

© Springer International Publishing AG 2016
S. Boumerdassi et al. (Eds.): MSPN 2016, LNCS 10026, pp. 183–196, 2016.
DOI: 10.1007/978-3-319-50463-6_15

the security necessary updates. Moreover, due to the fact that applications are executed in separate containers, the scope of many security applications is very limited. Additionally, most devices store a lot of data in storage media which are using file systems like FAT32, which do not support permissions allowing many malicious applications to perform further damages.

The "freemium" model has become the dominant business model on the Internet. The concept on which it is based is that users get the basic functionality for free, while other services and products can be purchased later. Since the service is free, users can easily access it and without monetary risk. Building upon the success and trust that these free services provide, a user can then appreciate the extended features that can be purchased.

While "premium" services is one of the income sources, the other main source is advertisement. Most of the companies operating under the "freemium" model monetize their wide user pool by displaying targeted advertisement or by supplying data to advertising companies. Based on the latter, thousands of start ups and developers are building their mobile applications and distribute them in Google Play, App Store etc. The prevalence of this model is questioned quite often with one of the major arguments being:

"If you're not paying for the product, you are the product."

Regardless of whether someone agrees with this statement or not, since the "freemium" model is widely adopted, the major question is whether this model implies any exposure for the user. One could argue that depending on the interaction, companies could profile users and the fusion with other data could lead to the disclosure of personal and sensitive information. A typical example is the location of a user. Solely it does not mean anything, however, when mixed with other context it can disclose very sensitive information. For instance, the presence of someone in a temple, in an area where a demonstration is taking place or to a hospital, can enable someone to determine her religious or political beliefs, or her medical condition accordingly. Nevertheless, to deduce this information, the user has agreed to share this information with the service provider, therefore, it can be considered a calculated risk. But is that really all her exposure? Does the implementation of the "freemium" model implies further user exposure?

AndroPatchApp tries to tackle many of these issues in Android, without the need to perform changes to the operating system or requesting elevated permissions. To achieve this, we go a step backwards, instead of making the changes to the devices, we enable the changes to be made in the application. Therefore, AndroPatchApp acts before the installation of an app. The user uploads to the desired application which is decompiled and some additional libraries are embedded before packaging it. This way, the user installs an application which is less pervasive and with some embedded controls to detect malicious activities.

The rest of this work is structured as follows. The next section introduces the reader to the problem, illustrating the to what risks a user is exposed to. Then, in Sect. 3, we provide an overview of the related work. In Sect. 4 we present our proposed solution, compare it to current state of the art and illustrate some experimental results. Finally, the article concludes summarizing our contributions.

2   Problem Setting

As already discussed, one of the main income sources in the "freemium" model are advertisements. To embed them in the apps there are several ways. In Android, the most common ad library is AdMob, Fig. 1, which allows apps to display ads in the header or the footer. In principle, this means that the developer adds a browser-like object; a webview, in his app which receives HTML and JavaScript code from the ad server and displays it accordingly.

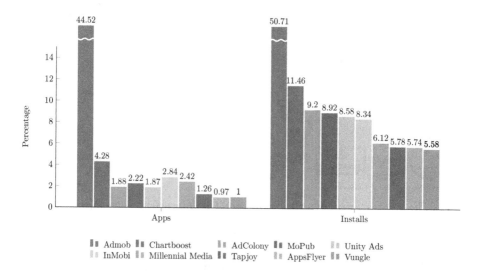

Fig. 1. Statics of Andoid ad libraries. Source: http://www.appbrain.com/ [Accessed: March 2016]. Note that an application may use more than one ad libraries simultaneously.

It has to be noted that the current model of advertisements works over HTTP. The use of HTTP facilitates ad servers as they can cache their content and compress it significantly increasing their throughput. However, since the content is delivered in plaintext, an adversary could easily perform remote code injection. Therefore, the use of HTTP exposes the user to attacks which may stem from the network administrator or provider (e.g. the case of Verizon "super-cookies" [1]); he changes the content of the delivered ads, or in more stealth attacks, other users of the network. These scenarios are illustrated in Fig. 2. The user is also exposed to attacks from the advertiser, as the ads he delivers to the ad server for distribution might be malicious [2].

The concept of ad injection is very simple, an adversary takes control of what is displayed in the browser or the app of his victims and injects his ads. This way, the advertiser receives legitimate traffic and can increase his target audience, so the advertiser is willing to pay the injector for his "services". While the injector and the advertiser have a "win-win" scenario, the quality of the

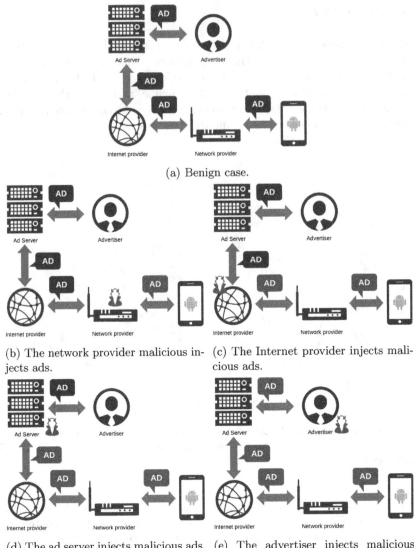

(a) Benign case.

(b) The network provider malicious injects ads.

(c) The Internet provider injects malicious ads.

(d) The ad server injects malicious ads.

(e) The advertiser injects malicious ads.

Fig. 2. Ad delivery and attack scenarios.

service that the user receives is degraded. The webpages and the apps are filled with nagging ads hiding the actual content from the user. This practice has become quite common, as for instance there is a specific type of malware, called adware, whose sole target is to inject ads in the victim's machine.

The extend that this trend has reached can be understood by the fact that 5.5% of the traffic of unique IPs studied by Google were injected with ads [3].

Taking ad injection to another level, we recently witnessed that specific models of a big manufacturer where sold with the notorious adware Superfish pre-installed [4]. In fact, Superfish is backed by a business which carries its name, which is so well that it was ranked 4th in the fastest-growing private companies in America, with revenue of $35.2 m in 2013 [5]. Additionally, hundreds of applications in Google Play have been found to be linked with tracking and ad related sites, and sites associated with malware activity [6], while others have been found to contain vulnerabilities [7].

Currently, Android is the most widely used platform for smartphones. The core of Android is Linux, therefore its core security mechanisms stem from it. Let us assume that a user wants to install an application to his device. Each application in Android is packaged in an APK file which is a compressed file containing everything that an application needs to be executed, such as code, icons and media. To install an app the Package Manager service invoked. Officially, *Google Play* is used to provide an easy way to manage applications. The other options is to invoke the Package Installer when downloading or coping an app to the phone and asking Android to open it, or using ADB. In the first two cases, the user will be presented with a screen which informs which permissions have to granted to the application in order to run in his device. Note that in all versions up to Marshmallow, the app permissions were given once to app and could not be revoked. However, in Marshmallow the user can revoke permissions or grant them upon request. While in some cases app permissions could be bypassed [8,9], in general they are considered to work efficiently. Notably, the continuous increase of app requests for access to sensitive permissions or unrelated to the application's functionality, have made users to ignore Android warnings [10]. These installation lifecycles are illustrated in Fig. 3.

Furthermore, the Android architecture enables the exchange of data with other services, applications and content providers through the inter-component communication (ICC) [11,12]. Thus, the developer can label the desired access level of each component within an application according to his taste by simply making the according declarations in the *AndroidManifest.xml*. The developer is also expected to define the permissions that the application needs in this file.

While the above architecture seems concrete, there are several problems. For instance, Grace et al. [13] found that many applications would use HTTP to download code and execute it on client's device. While insecure, this method is often seen in Android applications [14]. Nevertheless, there is another inherent threat: The access level to a resource is granted *per application* and *not per component*. Theoretically, this does not raise any important issue, since all the components are handled by the same entity, the developer who created the app.

Since most of the apps are currently working in the "freemium" model, developers include ads in their apps. To facilitate this task, most platforms include a specific component for this task. The developer simply provides an area in the app where the ads should be displayed and he links the component with his account to receive payments depending on how many times the users of his app have watched, clicked an ad. In the case of Android this is realized with the

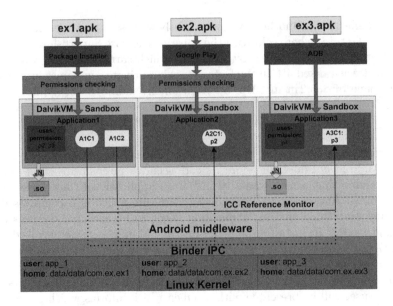

Fig. 3. App installation workflow. Source: [15].

use of a webview component, which is a small browser that displays HTML ads which are sent by the ad provider.

When a user decides to install an application, the first check that is performed is about is permissions. Should the user accept the requested permissions, the application is installed. Notably, only few users understand Android permissions [16,17]. Notably, in the latest version of Android 6, the user has more control of his apps, as he can selectively or permanently revoke a permission after the installation, when the app tries to make use of a granted permission. Nevertheless, the user cannot determine for which component of the app he is granting access. Apparently, once a permission is granted, every component of the app can use it. Therefore, the code that is executed from an ad in the provided webview has the same permission as the app itself. Regardless of whether the ad library supports geolocation, if the app has been granted location access, then the displayed ad has the same access as well. Note that this is information can become available not only to the ad provider, but to the ad server, the Internet or network provider, depending on who provided or injected the geolocation code.

Regardless of whether there is a code injection, ad libraries have not proven to be benign. For instance, Stevens et al. [18] found that some of them would use undocumented permissions, while Grace et al. [19] observed that approximately half of the ad libraries would probe the apps that contain them to determine whether they have more privileges and abuse them to derive sensitive user information.

Apparently, not only the libraries which are used for mobile ads can expose users' privacy, but the whole underlying architecture exposes users to further risks, allowing many entities to manipulate the ads and inject malicious code. Certainly, shifting to HTTPS could significantly increase the security of the apps. Nevertheless, this would only apply to outsider threats, someone who injects ads, and not the risks of the user to the ad library. Finally, HTTPS implementations in mobile devices have reportedly proven to be erroneous [20–22].

3 Related Work

In general, ads have started becoming more and more greedy about users' information. In fact, the more installations an app has, the more privacy invasive it tends to be [23]. Apart from requests to get user's location, ad libraries, may perform WiFi scans to determine users' location, scan whether the user has accounts in social networks or even scan the device to find which applications have been installed [24]. Recently, more advanced ad libraries manage to link devices by playing inaudible sounds from one device and collecting them from the microphone of mobile devices that use applications where such an ad library has been embedded [25].

Experimenting with ad libraries, Stevens et al. [18] found that some ad libraries would try to access of undocumented permissions such as read/write to calender or access location and camera. Moreover, since most of them used HTTP and Javascript, it was easy for an attacker to inject code in them and perform several attacks like exfiltrate and/or modify the user's calendar and contacts, exfiltrate user's audio and image files.

Gibler et al. [26] introduced a tool to scan for possible leaks of sensitive information named AndroidLeaks. This tool decompiles an APK to derive its code and generate a call graph. AndoidLeaks uses this graph to map the permissions and analyses through data flows whether private information could be leaked. The tool clearly can only be used to detect potential leaks and not prevent them, while it also does not analyses whether there is an actual leak; the app does not this information to be functional, or not; the app needs this information to work.

To counter problems from ads, several solutions have been proposed in the literature, such as AdDroid [27], or XPrivacy [28]. However, regardless of their effectiveness, they require rewriting the Android platform and installing it in the phone. TaintDroid [29] is a more generic tool to counter privacy leakage by monitoring information flow, however, it also requires rewriting of the Android platform.

Closely related to our work is AdSplit [30] of Shekhar et al. AdSplit decompiles apps to provide an augmented environment in which they are executed. The most important features of AdSplit are the distinct and independent permission sets of ads and applications as well as the separation into distinct processes. The distinct differences with of AndroPatchApp with AdSplit will be discussed in detail in the next section.

4   Proposed Solution

The scope of AndroPatchApp is to provide an alternative solution to the problem without the need for rooting the device or creating a new ROM and changing the current architecture of Android. Many of the aforementioned problems, specially in the case of exposure to ad libraries could be solved from the developers of ad libraries, by making them more privacy-aware and delivering the ads via HTTPS. Since the current model is very profitable for them, they have no incentive to change it. Additionally, rooting the device not only requires advanced knowledge from the users, but exposes them to further risks.

To break this loop, AndroPatchApp intervenes before the installation of the app, modifying the application to be installed by injecting smali libraries which can block several malicious requests. The workflow of AndroPatchApp is illustrated in Fig. 3 and it is the following: The user first uploads the apk file that he wants to patch to AndroPatchApp. On receiving the file, AndroPatchApp decompiles it using ApkTool [31] which returns the smali code of the app. Then, AndroPatchApp searches for all the instances of webviews in the code. Since all ad libraries are displayed through a webview, AndroPatch finds the webviews that belong to ad libraries and injects the according additional code, depending on user preferences. The goals of the injected code are the following:

Hide user's geolocation: As already discussed, ads regardless of whether they are legitimate or injected, they have access to the user's geolocation, as long as this access is granted to the app. To counter this issue, the injected code rewrites user's geolocation, returning a dummy value, in this case $(0, 0)$.

Obfuscate available resources: AndroPatchApp, can hide the available sensors, for instance the use of microphone or media. This is achieved by injecting a script which returs that the media and microphone are off. This approach disable nefarious practices of ad libraries such as "SilverPush".

Disable javascript: AndroPatchApp can allow only static content to be displayed or media, by simply disabling the javascript in all the webviews of ad libraries. This can be achieved by setting:

```
getSettings().setJavaScriptEnabled(false);
```

Remove content of an add: Finally, AndroPatchApp can totally remove the content of an ad, by simply injecting a code which sets the HTML code of the webview to null.

Since AndroPatchApp injects code only to the webviews, it does not change the data flows, dependencies of app or interfere with other parts of the code. Therefore, the functionality of the application is exactly the same as the original, therefore the repackaged application will not result to any crashes or display errors to the user, decreasing the user experience. Moreover, the whole procedure is automated and does not require supervision.

AndroPatchApp will then use ApkTool to repackage the application which is signed and returned to the user for installation.

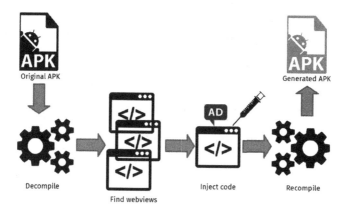

Fig. 4. AndroPatchApp workflow.

To inject the necessary code in the webviews, we take advantage of the *onPageFinished* method of webviews. This method notifies the app that the application has finished loading the code and that the corresponding HTML object has been created inside which we load our javascript. To remove the HTML content of an ad one can simply inject:

```
document.body.parentNode.innerHTML ="";
```

Additionally, one can simply change the webview behaviour by changing *setAllowUniversalAccessFromFileURLs*, *setAllowFileAccessFromFileURLs* and *setJavaScriptEnabled* from True to False, where appropriate. Note that all these methods are void, which facilitates the injections. For instance, AndroPatchApp traces in the decompiled smali code a call like:

```
invoke-virtual {v2, v6}, Landroid/webkit/WebSettings;->setJavaScriptEnabled(Z)V
```

and tries to find variable *v6* to assign it the value 0×0 before calling the method. So the code would be more or less like:

```
static_code_injection ='    const/4 '
control_webview_value_false=' 0x0'
```

```
1    myWebView.setWebViewClient(new WebViewClient() {
2
3    @Override
4    public void onPageFinished(WebView view, String url)
5    String js_loc = "var newscript = document.createElement
     (\"script\");";
6    js_loc = js_loc + "newscript.onload=function(){
     overrideLocation();};";
```

Listing 1.1. Overriding onPageFinished.

```
1   function overrideLocation () {
2       navigator.geolocation.getCurrentPosition = function(
    success, failure) {
3           success ({
4               coords: {
5                   latitude: 0,
6                   longitude: 0,
7               },
8               timestamp: Date.now()
9           });
10      }
11  }
```

Listing 1.2. Overriding location.

The apps, as illustrated in Fig. 5, continue working seamlessly, however, without ads.

Compared to AdSplit, AndroPatchApp has several additional features. Firstly, AndroPatchApp can be applied to almost any ad library with minimum effort, in fact the demo version of AndroPatchApp is applied not only to AdMob, but to a wide list of ad network libraries, for the complete list see Table 1. While AdSplit has to find all the API calls and replace them accordingly, AndroPatchApp simply injects the necessary code to the source of the problem: the WebView of the ads. Note that the former approach is quite sensitive to versioning and it is relatively easy to result to broken apps. Additionally, in order to provide the necessary functionality, AdSplit introduces a significant overhead on the application, whereas AndroPatchApp leaves the apps intact, therefore there is no additional memory or CPU overhead. Finally, and most importantly, AndroPatchApp is a standalone solution with no additional prerequisites and no modifications to stock Android installations. On the contrary, AdSplit depends on Quire [32] which needs a modification of the Android platform. Thus, AndroPatchApp is a more versalite approach as it demands the least possible user interaction and prerequisites.

As already discussed, MarshMallow allows users to revoke access to specific permissions or grant them upon request. While this is a significant improvement for users' privacy, it has to be noted that if an app needs access to sensitive information such as geolocation, then the user can either reject access to it, crippling the application of its core functionality, or grant it upon request. While the latter sounds fair, if the functionality is frequently used, then user experience is degraded with the constant requests. Finally, even if the user is prompted, he cannot tell whether the request is made for the app or for the ad library.

Table 1. Supported ad libraries

AdMob	Flurry	InMobi
TapJoy	MobClix	ChartBoost
AdWhirl	MoPub	GreyStripe

(a) Original app.   (b) Patched app.

(c) Original app.   (d) Patched app.

Fig. 5. Example screenshots of AndroPatchApp.

5 Conclusions

With the vast adoption of portable devices equipped with a plethora of sensors, ad networks have become more rogue. Exploiting the ability to extend their data collection capabilities to additional, and potentially sensitive, user information has dramatically increased the permission requests from apps that embed ad libraries. This trend has made users accustomed with such requests and greatly relaxed their privacy preferences. Nevertheless, this shift implies many risks for the users, not only in privacy, but in security as well.

In this work we introduced AndroPatchApp, a new tool which addresses this problem by sanitizing the APK files, without compromising the functionality of the app. More precisely, AndroPatchApp decompiles the installation files to derive the smali code and perform code injection to the webviews that are controled by ad libraries. This allows the user to customize the sanitization process, from removal of geolocation capabilities up to complete removal. All the above are achieved without compromising using experience, as AndroPatchApp does not introduce any computational overhead, while it does not require rooting the device.

Nevertheless, the biggest drawback of the proposed solution is the legal aspects. From one side we have the developer which might not want to allow others to decompile and process his code as these actions can decrease the income that he has from the ads. Nevertheless, one has to see the other side, that of the user. The user has not given his explicit consent to share sensitive information with other parties than the developer. In fact, such terms are hardly directly documented in the terms and conditions of a service or an app. If the user had the option to opt out of such conditions or he was aware that by using such an app would expose specific information to a third party it is quite possible that he would not choose it. Therefore, AndroPatchApp can be considered as a means to enforce methods to selectively choose what information is disclosed to whom and when on a platform where the user has constrained privileges.

Acknowledgments. This work was supported by the European Commission under the Horizon 2020 Programme (H2020), as part of the *OPERANDO* project (Grant Agreement no. 653704).

References

1. Hoffman-Andrews, J.: Verizon injecting perma-cookies to track mobile customers, bypassing privacy controls (2014). https://www.eff.org/deeplinks/2014/11/verizon-x-uidh
2. Mills, E.: Malware delivered by yahoo, fox, google ads (2010). http://www.cnet.com/news/malware-delivered-by-yahoo-fox-google-ads/
3. Thomas, K., Bursztein, E., Grier, C., Ho, G., Jagpal, N., Kapravelos, A., McCoy, D., Nappa, A., Paxson, V., Pearce, P., et al.: Ad injection at scale: assessing deceptive advertisement modifications. In: IEEE Symposium on Security and Privacy (SP), pp. 151–167. IEEE (2015)

4. Graham, R.: Extracting the superfish certificate (2015). http://blog.erratasec.com/2015/02/extracting-superfish-certificate.html
5. Inc. 5000 2015: The full list (2016). http://www.inc.com/inc5000/list/2014/
6. Vigneri, L., Chandrashekar, J., Pefkianakis, I., Heen, O.: Taming the android appstore: lightweight characterization of android applications, CoRR abs/1504.06093
7. Enck, W., Octeau, D., McDaniel, P., Chaudhuri, S.: A study of android application security. In: Proceedings of the 20th USENIX Conference on Security, p. 21. USENIX Association (2011)
8. Davi, L., Dmitrienko, A., Sadeghi, A.-R., Winandy, M.: Privilege escalation attacks on Android. In: Burmester, M., Tsudik, G., Magliveras, S., Ilić, I. (eds.) ISC 2010. LNCS, vol. 6531, pp. 346–360. Springer, Heidelberg (2011). doi:10.1007/978-3-642-18178-8_30
9. Orthacker, C., Teufl, P., Kraxberger, S., Lackner, G., Gissing, M., Marsalek, A., Leibetseder, J., Prevenhueber, O.: Android security permissions–can we trust them? In: Prasad, R., Farkas, K., Schmidt, A.U., Lioy, A., Russello, G., Luccio, F.L. (eds.) MobiSec 2011. LNICSSITE, vol. 94, pp. 40–51. Springer, Heidelberg (2012). doi:10.1007/978-3-642-30244-2_4
10. Felt, A.P., Ha, E., Egelman, S., Haney, A., Chin, E., Wagner, D.: Android permissions: user attention, comprehension, and behavior. In: Proceedings of the Eighth Symposium on Usable Privacy and Security, p. 3. ACM (2012)
11. Enck, W., Ongtang, M., McDaniel, P.: Understanding android security. IEEE Secur. Priv. 7(1), 50–57 (2009)
12. Felt, A.P., Chin, E., Hanna, S., Song, D., Wagner, D.: Android permissions demystified. In: Proceedings of the 18th ACM Conference on Computer and Communications Security, pp. 627–638. ACM (2011)
13. Grace, M.C., Zhou, Y., Wang, Z., Jiang, X.: Systematic detection of capability leaks in stock android smartphones. In: NDSS (2012)
14. Poeplau, S., Fratantonio, Y., Bianchi, A., Kruegel, C., Vigna, G.: Execute this! analyzing unsafe, malicious dynamic code loading in android applications. In: 21st Annual Network and Distributed System Security Symposium, NDSS, San Diego, California, USA, 23–26 February. The Internet Society (2014)
15. Zhauniarovich, Y.: Android security (and not) internals
16. Kelley, P.G., Consolvo, S., Cranor, L.F., Jung, J., Sadeh, N., Wetherall, D.: A conundrum of permissions: installing applications on an android smartphone. In: Blyth, J., Dietrich, S., Camp, L.J. (eds.) FC 2012. LNCS, vol. 7398, pp. 68–79. Springer, Heidelberg (2012). doi:10.1007/978-3-642-34638-5_6
17. Balebako, R., Jung, J., Lu, W., Cranor, L.F., Nguyen, C.: Little brothers watching you: raising awareness of data leaks on smartphones. In: Proceedings of the Ninth Symposium on Usable Privacy and Security, p. 12. ACM (2013)
18. Stevens, R., Gibler, C., Crussell, J., Erickson, J., Chen, H.: Investigating user privacy in android ad libraries. In: Proceedings of the Workshop on Mobile Security Technologies (MoST) (2012)
19. Grace, M.C., Zhou, W., Jiang, X., Sadeghi, A.-R.: Unsafe exposure analysis of mobile in-app. advertisements. In: Proceedings of the Fifth ACM Conference on Security and Privacy in Wireless and Mobile Networks, WISEC 2012, pp. 101–112. ACM (2012)
20. Fahl, S., Harbach, M., Muders, T., Baumgärtner, L., Freisleben, B., Smith, M.: Why Eve and Mallory love android: an analysis of android SSL (in) security. In: Proceedings of the ACM Conference on Computer and Communications Security, pp. 50–61. ACM (2012)

21. Conti, M., Dragoni, N., Gottardo, S.: MITHYS: mind the hand you shake-protecting mobile devices from SSL usage vulnerabilities. In: Accorsi, R., Ranise, S. (eds.) STM 2013. LNCS, vol. 8203, pp. 65–81. Springer, Heidelberg (2013). doi:10.1007/978-3-642-41098-7_5

22. Hubbard, J., Weimer, K., Chen, Y.: A study of SSL proxy attacks on android, iOS mobile applications. In: IEEE 11th Consumer Communications and Networking Conference (CCNC), pp. 86–91. IEEE (2014)

23. Book, T., Wallach, D.S.: A case of collusion: a study of the interface between ad libraries and their apps. In: Proceedings of the Third ACM Workshop on Security and Privacy in Smartphones and Mobile Devices, pp. 79–86. ACM (2013)

24. Book, T., Pridgen, A., Wallach, D.S.: Longitudinal analysis of android ad library permissions, arXiv preprint arXiv:1303.0857

25. Goodin, D.: Beware of ads that use inaudible sound to link your phone, tv, tablet, and pc (2015). http://arstechnica.com/tech-policy/2015/11/beware-of-ads-that-use-inaudible-sound-to-link-your-phone-tv-tablet-and-pc/

26. Gibler, C., Crussell, J., Erickson, J., Chen, H.: AndroidLeaks: automatically detecting potential privacy leaks in android applications on a large scale. In: Katzenbeisser, S., Weippl, E., Camp, L.J., Volkamer, M., Reiter, M., Zhang, X. (eds.) Trust 2012. LNCS, vol. 7344, pp. 291–307. Springer, Heidelberg (2012). doi:10.1007/978-3-642-30921-2_17

27. Pearce, P., Felt, A.P., Nunez, G., Wagner, D.: AdDroid: privilege separation for applications and advertisers in android. In: Proceedings of the 7th ACM Symposium on Information, Computer and Communications Security, pp. 71–72. ACM (2012)

28. Xposed module repository (2015). http://repo.xposed.info/module/biz.bokhorst.xprivacy

29. Enck, W., Gilbert, P., Han, S., Tendulkar, V., Chun, B.-G., Cox, L.P., Jung, J., McDaniel, P., Sheth, A.N.: Taintdroid: an information-flow tracking system for realtime privacy monitoring on smartphones. ACM Trans. Comput. Syst. (TOCS) **32**(2), 5 (2014)

30. Shekhar, S., Dietz, M., Wallach, D.S.: AdSplit: Separating smartphone advertising from applications. In: Presented as Part of the 21st USENIX Security Symposium (USENIX Security 12), pp. 553–567 (2012)

31. Winsniewski, R.: Android-apktool: a tool for reverse engineering android apk files (2012). http://ibotpeaches.github.io/Apktool/

32. Dietz, M., Shekhar, S., Pisetsky, Y., Shu, A., Wallach, D.S.: Quire: lightweight provenance for smart phone operating systems. In: Proceedings of the 20th USENIX Conference on Security, SEC 2011, p. 23. USENIX Association, Berkeley (2011)

Creating an Easy to Use and High Performance Parallel Platform on Multi-cores Networks

Viet Hai Ha[1(✉)], Xuan Huyen Do[2], Van Long Tran[3], and Éric Renault[3]

[1] College of Education, Hue University, Hué City, Vietnam
haviethai@gmail.com
[2] College of Sciences, Hue University, Hué City, Vietnam
doxuanhuyen@gmail.com
[3] SAMOVA, Télécom SudParis, CNRS, Université Paris-Saclay,
9 rue Charles Fourier, 91011 Evry Cedex, France
{van_long.tran,eric.renault}@telecom-sudparis.eu

Abstract. How to easily exploit the performance of network using multi-core processors nodes is the purpose of many researches including CAPE (Checkpointing Aided Parallel Execution). CAPE uses the checkpointing technique to bring the simplicity and high performance of OpenMP – a high performance and easy-to-use standard of parallel programming API on shared-memory architecture – onto distributed-memory architectures. Theoretical analysis and experimental results have proved that CAPE has ability of providing a high performance and complete compatibility with OpenMP standard. This article aims at introducing how to use multiple processes on calculating nodes to increase performance of CAPE with the initial results.

Keywords: CAPE · Checkpointing Aided Parallel Execution · OpenMP · Parallel programming · Distributed computing · HPC

1 Introduction

1.1 OpenMP

OpenMP [1] is an API providing a high level abstraction for parallel programming on shared-memory architectures. It consists of a set of environment variables, directives and functions that support easily converting sequential C/C++ or Fortran programs into parallel programs.

OpenMP uses fork-join model with thread as the basic parallel structure. Initially, the program consists of only one master thread processing sequence code. Whenever meeting an OpenMP parallel directive, this master thread spawns a team work including itself and a set of slave threads (phase fork) and tasks are divided into these threads. After the slave threads have finished their tasks, the result is updated in the memory space of the main master thread and they can finish work (phase join). So, after this phase, the program remains only one thread as the original program.

© Springer International Publishing AG 2016
S. Boumerdassi et al. (Eds.): MSPN 2016, LNCS 10026, pp. 197–207, 2016.
DOI: 10.1007/978-3-319-50463-6_16

OpenMP uses a relaxed-consistency, shared-memory model. All OpenMP threads have access to a place to store and to retrieve variables, called the memory. In addition, each thread is allowed to have its own temporary view of the memory. Currently, OpenMP has only been completely installed for the shared-memory architecture because of the complexity of the installation of all the requirements of OpenMP on the other memory model. This is the motivation for many researches to be conducted with the objective of installing OpenMP on the distributed-memory architecture. However, there is not any result having met the two requirements of fully compatibility with OpenMP standard and high performance. The most prominent alternatives may include the use of SSI [2]; SCASH [3]; compiled into MPI [4,5]; the use of Global Array [6]; and Cluster OpenMP [7]. Even with Cluster OpenMP, a commercial product from Intel also requires the use of its own more directives (not belonged in OpenMP standard) in some cases. And therefore, it is not a fully compatible installation of OpenMP.

1.2   CAPE

Principle of CAPE: CAPE (Checkpointing Aided Parallel Execution) [8] is a new approach to install OpenMP on distributed-memory systems. CAPE uses process as the basic parallel unit, instead of using thread in original OpenMP. With CAPE, all the most important tasks of the fork-join model are automatically implemented using checkpointing technique, including the division of task to slave processes, extraction of results slaves-ones and the updating these results in the memory space of the master process.

Deployment model: Figure 1 illustrates CAPE deployment diagram. In there:
Master node: plays the role of master thread in the operating model of OpenMP. Accordingly, it executes the code section of the master process, distributes the job to slave processes and receives the results achieved by slave processes after completing the parallel code blocks. These tasks are performed by the modules:

- User Application: the user's program code, originally written in the original language (CAPE is supporting with C language) along with OpenMP directives. This program has been translated by the CAPE program into a standard C code and then continues to be translated into machine code by a conventional C compiler, such as GCC of GNU.
- Distributor: This is the program which sets nodes in the system and distributes tasks for nodes. Distributor basing on IP of node to distinguish the nodes and activate the program on those nodes with the corresponding parameter. Currently, the distributor is installed by a Shell program, with the input parameters are the IP of the nodes in the system as well as the role of those nodes in the operational model of CAPE.
- Monitor: this program is both a checkpointer (snapshot progress of process) and also a management application program. In checkpointer role, it is a discontinuous incremental checkpointing [9], take on two main tasks are:

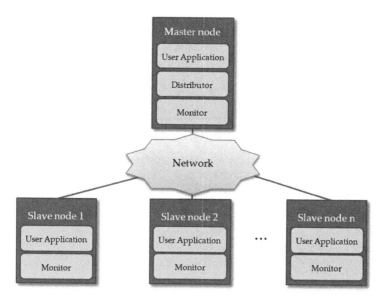

Fig. 1. CAFE deployment model

(1) make the snapshot progress in each specific code to initialize the status for the processes at the slave nodes at the beginning point of the parallel code; and (2) extract the implementation results of the parallel sections of code in slave nodes. Absolutely, it is also responsible for updating the memory space of the process by the snapshot processes which increase discretely at the identified locations. In the role of managing application program, the monitor is responsible for initializing application program; communicating and exchanging data between nodes including allocating the tasks, and sending calculation results from the slave nodes to the master one, etc.

Slave nodes: are the nodes performing the calculations in the parallel code. At those nodes, only two modules are the user's applications, which are required to play the role of calculation; and monitor. The operating model at these nodes is generating application program of users according to the requests sent from the distribution of master node, receiving calculation requirements, performing calculations and extracting calculation results back to the master node.

Results: CAPE has been developed and installed to achieve the parallel OpenMP directives such as parallel for, parallel section. We have carried out experiments on a square matrix multiplication, with various sizes from 3000×3000 to 12000×12000, for calculating the number of nodes varies between 2 and 30 [10]. Figure 2 shows the acceleration coefficient (speedup) of CAPE over the number of machines and different matrix sizes. The dashed lines represent the theoretical maximum increase. This chart shows clearly that the solution CAPE achieved very good results, with the measured speed ratio is in the range

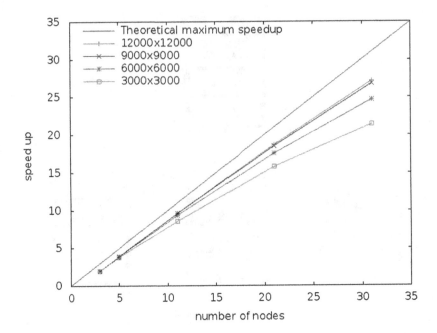

Fig. 2. Speedup of CAPE

of 75 to 90% in compare with the increase in the theoretical maximum acceleration coefficient. To some extent, the graph also performs the scalable (scalable)
of CAPE when the speed line is almost linear, and no sign of a significant decline
in the knotty calculations.

Disadvantages on Multicore Systems: Multi-core processor is a single
processor with two or more independent processing units, each unit can independently execute commands of programs. Nowadays, all new computers have
2 to 8 core processors; it leads to the popularity of computer networks with
multicore nodes. In order to reduce calculation time, application programs can
exploit the capabilities of multicore processors by running different parallel parts
of programs on these cores. OpenMP is a typical example of this direction by
using multithread execution model as mentioned in Sect. 1.1.

With the current execution model of CAPE, in each calculation node, there is
only one process running application programs. Moreover, this is a single process
and is not divided into sub-threads. Therefore, at each calculation node, the commands of the application programs are sequentially executed. This can be clearly
seen when looking at the graph measuring the execution parameters of system
while running a CAPE program, as shown in Fig. 3. As seen on this chart, only the
third core is fully exploited, while the other core is nearly inactive. Thus, the calculation resources of the system are being wasted and effectively exploiting them can
reduce the running time of programs, i.e. increase the efficiency of their executions.

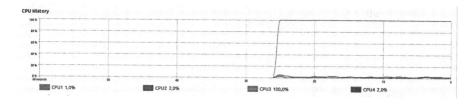

Fig. 3. Ratio of exploiting cores while running one process on each calculation node

2 Using Multiple Processes in Calculating Nodes to Increase Performance of CAPE in Multi-cores Systems

2.1 Principle

It should be noted that according to the execution model of CAPE, at the slave nodes, application code are sub-parts of parallel sections in original OpenMP programs. As consequence, these codes are potential in parallelization. Therefore, the general direction to effectively exploit the calculation resources in slave nodes is doing two things: (1) continue to divide the task at each node into subcomponents; and (2) make them concurrently running on different cores of processor. With multi-task operating systems today, the application program is not necessary to implement the second explicitly; it is taken over by the operating system. Meanwhile, with multi-process and multi-thread programs that use many calculation resources, the operating system will distribute calculating resources in an optimal way. For computers using multi-cores processors, the cores normally are load balancing. Thus, in the case of CAPE, only the first task has to be performed, by organizing the code in multi-process or multi-thread models. In there, multi-process model can be implemented directly by running multiple applications on a slave node, i.e. multiple processes on one physical machine. In this way, for each running time, the program will create a process and it will take a calculation part of parallel code.

2.2 Implementation Details

To create multiple processes by running multiple applications on a slave node, it is necessary to modify the distributor in the CAPE deployment model so that the application program is initialized many times with different parameters. Therefore, in each slave node, there are many application processes that performed different parts of the parallel sections. This is easily done as shown in the example below, in which the system has three nodes including one master node with IP address is 192.168.122.1 and two slave nodes with IP address respectively is 192.168.122.179, and 192.168.122.223. The name of application program is mulmt.

The original code of the distributor is a Shell script, when only one process of application program on each slave node is shown below (in which the ordinal number is added for the convenience of the presentation).

```
1.  #!bin/sh
2.  folder=/home/hahai/cape2/cdv9
3.  prog=mulmt
4.  num_nodes=2
5.  master=192.168.122.1
6.  node1=192.168.122.179
7.  node2=192.168.122.223
8. ${folder}/dbpf -f ${folder}/${prog} -a ${master}
    -k ${num_nodes} -o 0
9. ssh ${node1} ${folder}/dbpf -f ${folder}/${prog}
    -a ${master} -k ${num_nodes} -o 1 &
10. ssh ${node2} ${folder}/dbpf -f ${folder}/${prog}
    -a ${master} -k ${num_nodes} -o 2 &
11. exit 0
```

Explanation:

- Line 1: Specify interpreter Shell sh will be used
- Line 2: Specify the location of CAPE program
- Line 3: Specify the name of application program
- Line 4: Number of slave nodes
- Line 5: IP of the master nodes
- Line 6,7: IP of the slave nodes
- Line 8: Initialize the monitor and the application program on the master node
- Line 9,10: Initialize the monitor and the application program on the first and the second slave nodes
- Line 11: Exit Program

With the command lines above, the system will be initialized when the user execute this shell programs at the master node. Meanwhile, by the command at line 8, the monitor (dbpf) will start and it enables the application program (mulmt). The parameters of master IP address, number of slave nodes, and indicators of process are also transferred. Because the process index transferred is 0 (parameter - o 0), the application program should know it will act as the master node in the execution model of CAPE. The command lines number 9 and 10, are executed by the remote invocation ssh with IP address of the slave node. The other parameters are the same as in the command line number 8, except the process index is not 0 so that the application may know it is a slave node in the execution CAPE model. It can be noticed that with each slave node, there is only one call to the application program so there is only one application process is initialized. Moreover, when analyzing the conversion model of CAPE serves for transferring the OpenMP parallel constructs, it can be seen that in the slave nodes the application code is sequentially performed. Therefore, at each slave node, there is only one single process of application programs. This makes CAPE can not fully exploit the capabilities of multi-core processors, as stated in Sect. 1.2. In order to overcome this drawback in the way of running many times the application program on each slave node, the command lines number 9 and

10 will be cloned as shown in the code below, in which the application program is run twice on each slave node. Finally, the code of distributor is rewritten with the lines number 4 and 9, 10 are modified as below, while the other lines remain as original codes.

```
4.   num_nodes=4
...
9.   ssh ${node1} ${folder}/dbpf -f ${folder}/${prog}
     -a ${master} -k ${num_nodes} -o 1 &
9a.  ssh ${node1} ${folder}/dbpf -f ${folder}/${prog}
     -a ${master} -k ${num_nodes} -o 2 &
10.  ssh ${node2} ${folder}/dbpf -f ${folder}/${prog}
     -a ${master} -k ${num_nodes} -o 3&
10a.ssh ${node2} ${folder}/dbpf -f ${folder}/${prog}
     -a ${master} -k ${num_nodes} -o 4 &
```

In which:

- Line 4: is modified to have 4 slave processes.
- Line 9: is cloned into lines 9 and 9a, with the process index in line 9a is 2. Thus, in the first slave node, there is 2 times the application program is executed, i.e. there are 2 processes are created for the parallel code.
- Line 10: is processed in the same way of line 9, which is cloned into line 10 and line 10a, with process index is 3 and 4 respectively.

3   Experiments

To evaluate the feasibility as well as the performance of the proposed method, we have tested it with the matrix multiplication problem on a cluster with nodes equipped an Intel Core i3 processors (2 cores - 4 threads) running at 3.5 GHz, 4 GB RAM, uses Ubuntu 14:04, connected by 100 Mb/s Ethernet. The experiments were conducted with two scenarios are varying number of processes on the slave nodes and the matrix sizes. Some of the test results are presented below.

3.1   Taking Advantages of Multi-core Processors

Due to many application processes with high requirement of calculation resources running in concurrent, the capacity of multi-core processors are exploited better. This is clearly shown in the graph the proportion of the execution of the processors. Such as the case of 4 processes running on each slave node, the ratio of the core activities reaches 100%, as seen on the chart in Fig. 4. These results are very different from the case in which there is only one application process on each slave node, when only one core is exploited with its full capacity, while the other ones are nearly inactive, as shown in Fig. 3. All proved that the use of multiple processes has better exploited the performance of multi-core processors.

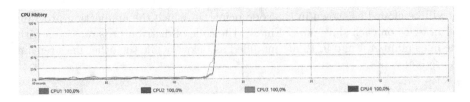

Fig. 4. Ratio of exploiting cores while running multiple processes on each slave node

3.2 Acceleration with Different Processes on Each Slave Node

To evaluate this, we have tested on a system with 11 nodes (1 master node, and 10 slave nodes); with the matrix size is 6000 × 6000. The experiment is conducted with the number of application processes on each slave node respectively is 1, 2 and 4. Note that the case of running one application process is also the case with the old execution model of CAPE.

The chart shows that the execution time decreases when running multiple processes. For the case of running 2 processes, execution time is reduced to nearly a half in comparing with the case running one process. This is reasonable cause for each node at this section, the number of calculation commands that are the ones that consumes the most calculation resources, are reduced by a half. For the case of running 4 processes, the processing time is decreased but it is also greater than the case of 2 processes. This can be explained by the mechanism of CAPE, whenever executing an application process, it is always accompanied by a process of monitor. Therefore, in fact, when running 4 processes of application program, there are up to 8 processes implemented in parallel. Although monitor process does not take too much calculation resources, it also has a certain influence on the distribution of system resources.

On the side of the master node, the result is good while running two processes in which the execution time also reduced by approximately a half. However, in the case of using 4 processes, the time strongly increases, even higher than the case of using one process. This result is unreasonable if we don't analyze it carefully. Back to the results in Fig. 2, when using an application process on each slave node, CAPE can generate a nearly linear speedup with the number of nodes, i.e. the number of calculation processes, including the maximum number of calculation nodes is 30. Consequently, there are two main causes of the abnormal increase of the execution time. The first is the architecture of the processors, with 2 real cores and 4 threads, instead of 4 real cores. The second is due to the multiple processes in the slave nodes have overlapping IP addresses when processing requests of setting the socket from these nodes, that causes a conflict and this needs time for solving. This is also the cause of the system failure when increasing the number of the processes on each slave node. However, this conclusion needs to be tested by conducting experiments and measurements in more details.

As shown in the Fig. 5, it is also the preliminary conclusion that with the machines using 2 cores – 4 threads, the optimal number of processes is 2.

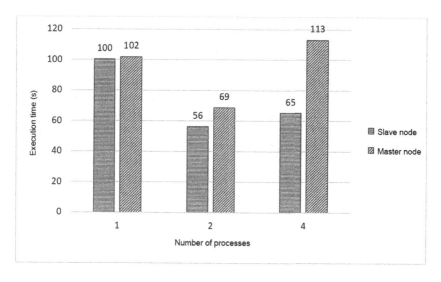

Fig. 5. Compare the execution time of program with different number of processes

3.3 Acceleration with Different Problem Sizes

To evaluate the scalability according to the problem sizes of multiple processes model, a similar experiment was conducted, with 6 nodes (5 slave nodes and 1 master node), using 4 processes on each slave node. The results of measurement are shown in the diagram in Fig. 6. These results are accordance with the complexity of matrix multiply problem.

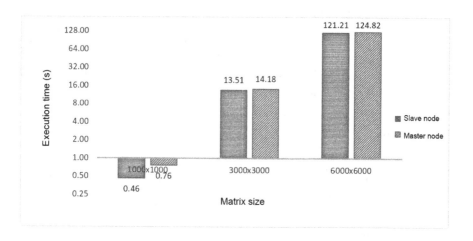

Fig. 6. Acceleration with different problem sizes

3.4 Advantages and Disadvantages

The outstanding advantage of this approach is simplicity, the program codes are almost unchanged (except for editing in the distributor). The experiments showed the performance is significantly increased, nearly double in the cases using 2 processes on each slave node. Theoretically, the slave nodes using N-core processors can run N times the application program to maximum exploit the processing capabilities. However, each application process requires one monitor process so the number of application processes in optimal case is less, and in this experiment, is N/2. This is also the first drawback of this direction. The second drawback is the possibility of conflicts among the programs on the same node. Moreover, the use of socket to implement the data transmission on network can also cause conflicts over resources between the programs and cause error with the large processes numbers. Finally, the use of multiple independent processes of the same program requires the slave nodes to have big amount of RAM to ensure the speed of execution.

4 Conclusion

CAFE, with the basic principles and the initial experimental results have shown its great potential to become a fully compatible implementation and high performance for OpenMP on distributed-memory architecture. Some researches are being continued to develop CAPE in many directions towards a fully implementation of OpenMP on these architectures, as well as in the direction of exploiting the new architecture of the processor to increase its performance. For networks using multi-core processors machines, the direction of using multiple processes on the calculation nodes by running multiple application programs has been tested and presented in this paper. This is a simple way, without requiring many changes in CAPE programs and but provides higher performance than the case of running one process in the previous model. However, there are still many shortcomings that need to be overcome such as resource conflicts which reduces the performance and causes system errors. That is one of our directions for developing CAPE in the near future.

References

1. OpenMP API: OpenMP application programming interface 4.5 (2015)
2. Morin, C., Lottiaux, R., Vallée, G., Gallard, P., Utard, G., Badrinath, R., Rilling, L.: Kerrighed: a single system image cluster operating system for high performance computing. In: Kosch, H., Böszörményi, L., Hellwagner, H. (eds.) Euro-Par 2003. LNCS, vol. 2790, pp. 1291–1294. Springer, Heidelberg (2003). doi:10.1007/978-3-540-45209-6_175
3. Sato, M., Harada, H., Hasegawa, A., Ishikawa, Y.: Cluster-enabled OpenMP: an OpenMP compiler for the SCASH software distributed shared memory system. Sci. Program. **9**(2, 3), 123–130 (2001)

4. Basumallik, A., Eigenmann, R.: Towards automatic translation of OpenMP to MPI. In: Proceedings of 19th Annual International Conference on Supercomputing, pp. 189–198. ACM (2005)
5. Dorta, A.J., Badía, J.M., Quintana, E.S., de Sande, F.: Implementing OpenMP for clusters on top of MPI. In: Martino, B., Kranzlmüller, D., Dongarra, J. (eds.) EuroPVM/MPI 2005. LNCS, vol. 3666, pp. 148–155. Springer, Heidelberg (2005). doi:10.1007/11557265_22
6. Huang, L., Chapman, B., Liu, Z.: Towards a more efficient implementation of OpenMP for clusters via translation to global arrays. Parallel Comput. **31**(10), 1114–1139 (2005)
7. Hoeflinger, J.P.: Extending OpenMP to clusters. White paper (2006)
8. Renault, É.: Distributed implementation of OpenMP based on checkpointing aided parallel execution. In: Chapman, B., Zheng, W., Gao, G.R., Sato, M., Ayguadé, E., Wang, D. (eds.) IWOMP 2007. LNCS, vol. 4935, pp. 195–206. Springer, Heidelberg (2008). doi:10.1007/978-3-540-69303-1_22
9. Ha, V.H., Renault, É.: Discontinuous incremental: a new approach towards extremely lightweight checkpoints. In: 2011 International Symposium on Computer Networks and Distributed Systems (CNDS), pp. 227–232, IEEE (2011)
10. Ha, V.H., Renault, E.: Improving performance of CAPE using discontinuous incremental checkpointing. In: 2011 IEEE 13th International Conference on High Performance Computing and Communications (HPCC), pp. 802–807, IEEE (2011)

Performance Evaluation of Peer-to-Peer Structured Overlays over Mobile Ad Hoc Networks Having Low Dynamism

Anurag Sewak[✉], Mayank Pandey, and Manoj Madhava Gore

Computer Science and Engineering Department,
Motilal Nehru National Institute of Technology Allahabad, Allahabad, India
{anuragsewak_2013rcs51,mayankpandey,gore}@mnnit.ac.in

Abstract. To exploit the synergy between Mobile Ad hoc Networks (MANETs) and structured Peer-to-Peer (P2P) overlays, three different approaches, namely cross-layered, integrated, and layered have been proposed for their integration. In the cross-layered approach, information from the lower layers is made available at the application layer or vice versa. The integrated approach implements peer-to-peer routing algorithms at the network layer or MAC layer, whereas the layered approach utilizes application layer virtual links to construct P2P overlay on top of the MANET routing layer. It has been observed that cross-layered approach and integrated approach perform relatively better in situations where both node mobility and churn rate are high, but these approaches require changes in the lower layers of the protocol stack. On the other hand, the layered approach does not require any change in the protocol stack at lower levels, thus facilitating easy implementation. However, under highly dynamic scenarios this approach fails to achieve satisfactory performance. This paper targets to analyze the performance of layered approach under low dynamism. We have evaluated the performance of two topologically different structured overlays utilizing the layered approach in scenarios where, mobility of nodes is minimal, churn rate is low and network size is limited. The performance evaluation has been done using different network sizes, mobility speeds and lifetimes of nodes. Further, the frequency of execution of overlay maintenance procedures is also varied according to the dynamics of the MANET. The simulation results are gathered and compared with respect to packet delivery ratio, latency and maintenance overhead. The results obtained establish the applicability of layered approach for implementation of structured overlays over MANETs having low dynamism.

Keywords: P2P · Structured overlay · MANET · Mobility · Churn · Dynamism

1 Introduction

Mobile Ad hoc Networks (MANETs) are infrastructure-less networks where nodes communicate among each other using multi-hop wireless links. The participating nodes in these types of networks act as sender, receiver or router

© Springer International Publishing AG 2016
S. Boumerdassi et al. (Eds.): MSPN 2016, LNCS 10026, pp. 208–223, 2016.
DOI: 10.1007/978-3-319-50463-6_17

depending upon the context. The movement of nodes in these networks makes the packet forwarding difficult due to continuously changing topology. To handle this, many routing protocols have been proposed such as Ad hoc On Demand Distance Vector Routing (AODV) [23], Dynamic Source Routing (DSR) [18], Optimized Link State Routing (OLSR) [17], etc.

In the last few years, Peer-to-Peer (P2P) overlays became very popular for direct communication among nodes in wired networks. In this technology also, participating nodes can act as sender, receiver or router. These nodes exchange data using transport layer end-to-end virtual links, where the data actually flows through the physical underlying network. The P2P overlay networks can be broadly classified as unstructured overlays (Gnutella [26], BitTorrent [24], etc.) and structured overlays (Chord [30], Pastry [27], Kademlia [20], etc.).

There exists a synergy between these two types of networks in terms of their design goals and principles [15]. They share many key characteristics such as self-organization, decentralization and dynamic topology. These common characteristics lead to further similarities between the two types of networks: (1) both have a flat and dynamically changing topology, caused by node join and node leave operations, and mobility of nodes in MANETs; and (2) both follow hop-by-hop connection establishment for communication. In P2P networks, these connections are established via TCP links with geographically unlimited range, whereas connections in MANETs are established via wireless links, limited by the radio transmission range.

Three approaches (cross-layered, integrated and layered) are being used by the researchers to combine the two paradigms [9]. The cross-layered approach facilitates exchange of information between the application, network, and MAC layers. The integrated approach brings down the peer-to-peer routing concepts at the network layer. The layered approach simply deploys the P2P overlay on top of the MANET routing layer. The cross-layered approach and integrated approach give acceptable performance in most of the situations, when used with MANETs having moderate mobility and churn patterns. But, the implementation of these approaches is quite difficult, as they demand modification in the lower layers (network and MAC layers) of the protocol stack, which is a tedious task [2, 9]. The complexity of the architecture in such implementations is also high, due to added functionalities at different layers of the protocol stack. However, the layered approach has not been exploited much by the researchers for deploying P2P overlays over MANETs, due to mismatch between the dynamics of the underlay network and the logical overlay. High dynamics of the mobile nodes in the MANET underlay makes it difficult for the overlay application to adapt the changes and adjust its structure and logical links [10, 12, 21]. Also, the overhead of exchanging maintenance messages increases tremendously and it becomes really difficult to scale up to large networks with such deployment [9].

It has been observed that, at places like airports, railway stations, hotels, conference halls, university campuses, workplaces, etc., people generally carry mobile devices like smart phones, tablets or laptops with them. These places have very low mobility or almost static conditions; however the movement pattern is usually random. Also, the rate at which people add to the scenario or leave the scenario, i.e. churn is moderately low, but the pattern is random. In such scenarios, people may be interested in sharing content of similar interest (documents, images, music, videos, etc.) stored on their mobile devices with each other. These devices can form a mobile ad hoc network by connecting via their IEEE 802.11 wireless interfaces in ad hoc mode, and run peer-to-peer file sharing overlay applications to share content. Structured P2P overlay applications like Chord [30], Pastry [27], Kademlia [20], etc. are well suited for sharing content between these devices in such scenarios. These applications maintain a tight control over the network topology and use a Distributed Hash Table (DHT) to route search queries for locating the desired content at specific nodes in the network. Due to low dynamics (low mobility and low churn rate) of mobile devices in such scenarios, the layered approach can suitably be adapted to run structured P2P overlay applications on top of the MANET routing layer. Such deployment is expected to give satisfactory performance, as the adaptability of the physical underlay network by the logical overlay is minimal, and no changes are required to be made in the protocol stack.

In this paper, we have taken two structured overlay protocols having different overlay topologies, ring-based Chord (the most cited one) and tree-based Kademlia (the most deployed one), and have used layered approach to test their performances over MANETs with low mobility and low churn conditions. The effects of churn conditions on these structured overlays have been studied in [32], where the mobility of nodes was not considered, as the study was conducted for fixed networks. Similar studies based on churn dynamics for structured P2P overlays have been carried out in [4,5,19,31]. Here, in this paper we have carried out the performance study for MANETs, which have dynamism in terms of both churn and mobility. To model the churn dynamics of the overlay nodes, we have taken a churn model which uses a probability distribution for generating lifetimes of nodes. The mobility pattern of nodes is modelled by a random mobility model. The performance evaluation of these structured overlays over MANET has been done using parameters like, network size, mobility speed of nodes, lifetime of nodes and overlay maintenance interval. Here, we have not considered the network partitioning and merging aspects of P2P overlays running over separate MANET clusters.

The remainder of this paper is structured as follows: In Sect. 2, we discuss the approaches currently being used for deploying P2P overlays over MANETs, along with their merits and demerits. The complete simulation setup is given in Sect. 3. The simulation results of deploying Chord and Kademlia over MANET under different parametric variations are discussed in Sect. 4. The conclusion of the work and future research directions are given in Sect. 5.

2   P2P Deployment Approaches over MANET

Several attempts have been made in the past to deploy unstructured and structured P2P overlays over mobile ad hoc networks using different approaches. But, more emphasis is on deploying structured P2P networks over these multi-hop networks, as unstructured P2P networks suffer from significant communication overhead due to their flooding and random walk nature of message propagation [9]. Deploying structured P2P overlays over MANETs help in reducing the communication overhead, as the DHT concept is pushed to the ad-hoc routing layer itself, enabling key-based routing for MANETs [2,9]. In some cases, physical proximity of nodes is also considered while building the overlay, in order to reduce the overhead [2]. The deployment of DHT-based structured P2P overlays over MANETs can possibly be done using any of the following approaches - cross-layered approach, integrated approach or layered approach [9].

Cross-layered Approach. In the cross-layered approach, communication between peer-to-peer layer and the underlay layers (MAC layer and network layer) is established in order to improve the performance. CrossROAD [13], MeshChord [6] and Ekta [25] are some cross-layered solutions for deployment of structured P2P overlays over MANETs. These cross-layered solutions offer reduced message overhead and improved response time, but developing cross-layer interactions between the P2P layer and the lower layers demand change in the protocol stack [2,9].

Integrated Approach. In the integrated approach, the peer-to-peer layer is integrated with the routing layer beyond the strict layering rule, where DHT concepts are implemented at the routing layer itself. Virtual Ring Routing (VRR) [7], Scalable Source Routing (SSR) [14], MADPastry [33], M-DART [8], Hashline [29], etc. have used the integrated approach for fusion of P2P overlay with MANET underlay. These solutions also offer less maintenance overhead, but implementing DHT at the routing layer requires significant changes in the protocol stack [2,9].

Layered Approach. In the layered approach, the P2P overlay is deployed "as-is" on top of the MANET routing layer. Very few attempts have been made to exploit the layered approach for deployment of structured P2P over MANETs. Bamboo/AODV [10] used the layered approach, where Bamboo DHT was deployed over a multi-hop wireless network running AODV. Proactive overlay management used by Bamboo resulted in increase of network overhead, but succeeded in maintaining the correctness of the network structure by providing efficient queries along the overlay. In [12,21], Chord was deployed over MANET using the layered approach, where its performance was evaluated on the basis of varying parameters, like network size, mobility speed, offered application load and application dynamics. In all the test cases, the network was inconsistent and lookups were not resolved properly, resulting in incorrect application behaviour. A performance comparison of various DHT based structured overlays over MANETs under high churn conditions can also be found in [11]. These proposals show that the deployment of a P2P protocol "as-is" on top of ad-hoc routing layer causes significant message overhead and redundancy in communication [10–12,21]. But, none of them have tested the applicability of the layered approach to deploy structured P2P overlay on top of MANETs under low mobility and low churn conditions.

The literature on structured P2P overlays in MANETs predominantly argues that a number of issues arise in adapting the logical overlay to the underlying physical wireless network. These issues are bandwidth constraints, overlay maintenance, network recovery mechanisms, multi-hop routing stretch, efficient query propagation and content distribution [9]. Some approaches given in the literature advocate for topology adaptation in order to handle the mismatch between the logical topology of the P2P overlay and the physical topology of the MANET underlay [1,28]. Topology awareness in cross-layered and integrated approaches helps in uplifting the performance, but requires significant change at the lower layers. In [28], an adaptive approach is used to build an unstructured overlay over MANET. Proximity aware approaches for structured overlays over MANETs, like 3D-RP [1] have also been proposed, that uses a 3D-space to consider physical intra-neighbour relationship of peers and exploits a 3D-overlay to interpret that relationship. This approach is suitable for MANETs with low mobility, but incurs an extra overhead for using the 3D-space to create intra-neighbour relationships [2].

3   Simulation Setup

In this paper, we have evaluated the performance of two structured overlays having different logical structures - Chord (ring-based) and Kademlia (tree-based), over different MANET underlay environments with varying parameters like network size, node mobility speed, lifetime of nodes and overlay maintenance interval. A detailed description of Chord and Kademlia can be found in [20,30] respectively.

3.1   Simulation Design

The simulation has been performed using the peer-to-peer simulation platform, OverSim [3] which runs over an integrated simulation environment comprising of OMNeT++ 4.6 [22] and INETMANET 2.0 Framework [16]. Our simulation model is based on the layered approach of implementing overlays on top of MANETs, Fig. 1. The overlay protocols, Chord and Kademlia run over the MANET routing layer and perform key-based routing using a test application at the peer-to-peer layer. AODV has been considered as the multi-hop routing protocol for our MANET underlay design. All nodes in the underlay adhere to IEEE 802.11 standard at the MAC layer, and the physical interface is provided by wireless radio.

Overlay. The overlay application running on each terminal contains a UDP module that generates UDP messages for the peer; an Overlay module that contains essential modules of Chord or Kademlia protocols; and a KBR (Key-based Routing) Test Application module used to test the routing of messages to peers. KBR Test Application is used to perform one-way test by exchanging PING messages between the overlay nodes. On successful receipt of PING messages, PONG messages are returned as acknowledgment back to the sender. Iterative

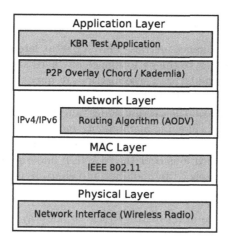

Fig. 1. P2P overlay over MANET - layered approach

routing was used to forward Chord/Kademlia and KBR Test Application messages from one overlay node to another. In this routing scheme, the source node starts a lookup by picking the closest next-hop node from its local routing table and sends the query message to obtain better next-hop nodes. These next-hop nodes are now successively queried by the source, till the closest sibling node is found.

Underlay. The underlay network represents the network topology which corresponds to the physical, MAC and network layers of the layered model. None of the underlay network models [3] available in the OverSim simulation platform (Simple Underlay Network, Single Host Underlay Network or INET Underlay Network) can be directly utilized for creating a network of wireless ad hoc nodes and implementing MANET routing. After analyzing all the possibilities of modification in the OverSim framework, we have done significant changes in the Simple Underlay Network model for deploying overlay protocols over a mobile ad hoc network of wireless devices. We have modified the Simple Underlay Network by creating a new Network Description (NED) file [22] for the ad hoc terminal with following specifications, in order to generate the desired underlay of wireless ad hoc nodes -

1. An IEEE 802.11 wireless interface was created for each terminal at the MAC layer.
2. A wireless radio interface was created for each terminal at the physical layer.
3. *Random Waypoint Mobility* model was imported from the INETMANET framework to define the mobility model.
4. AODVUU (an implementation of AODV developed by Uppsala University, Sweden) was defined as the MANET routing protocol for multi-hop routing at the network layer.

3.2   Performance Metrics

We have used the performance metrics - delivery ratio, latency and overlay maintenance overhead, to analyze the performance of Chord and Kademlia protocols over the MANET routing layer. These metrics are important, as they show the share of successfully acknowledged test messages, the time taken by the test messages to travel between the overlay nodes, and the overhead incurred by the overlay application on exchanging maintenance messages between overlay nodes, respectively. These metrics are relevant for performance evaluation of the two structured overlay applications over the MANET underlay having dynamically changing topology (influenced by mobility and churn dynamics). These metrics can be defined as follows:

- **Delivery Ratio:** The mean delivery ratio of one-way test messages sent by the KBR Test Application running on each overlay terminal. The delivery ratio represents the ratio of the number of messages received at the receiver to the number of messages sent by the sender, for each message sent/received.
- **Latency:** The mean latency of one-way test messages sent by the KBR Test Application running on each overlay terminal. One-way latency measures the time delay between the transmission time of a message at the sender and the arrival time of that message at the receiver, for each message sent.
- **Overlay Maintenance Overhead:** The mean number of overlay maintenance messages sent per second by the base overlay application. The maintenance of overlay application is a regular feature which is an overhead measured in terms of mean number of maintenance messages or bytes exchanged between overlay nodes per unit time.

3.3   Simulation Parameters

Certain parameters of the simulation scenarios supplied to the configuration files of OverSim, *omnetpp.ini* and *default.ini*, were uniform across all configurations, and are given in Table 1.

Each configuration was iterated ten times, where the total number of configurations in each simulation setting were twelve (four variations of network size, with three variations either of mobility speed, lifetime, or overlay maintenance interval). The network size determined by number of target overlay terminals was varied from 10 to 200 nodes in each simulation run. The transmission power of nodes was set to 32 mW (IEEE 802.11 standard for laptops). The mobility and churn patterns were chosen as per the nature and dynamics of the application scenario. The performance results were recorded at low mobility speeds randomly generated between 0.1 m/s and 2 m/s using the *Random Waypoint Mobility* model. The churn rate was controlled by the *Lifetime Churn* model of OverSim, in which nodes were added to the scenario after a deadtime drawn from a probability distribution function, and removed from the scenario on expiry of their lifetimes generated from the same probability function.

Table 1. Uniform simulation settings for all configurations

Simulation parameter	Value	Module
Simulation Area	1000 m × 1000 m	Overlay Terminal
Initial X, Y positions	10 m, 10 m	Overlay Terminal
Init Phase Creation Interval	0.1 s	Churn Generator
Tier 1 Type	KBR Test App	KBR Test Application
Underlay Type	Simple Underlay	Underlay Configurator
Terminal Type	Simple Overlay Ad hoc Host	Underlay Configurator
Graceful Leave Delay	15 s	Underlay Configurator
Graceful Leave Probability	0.5	Underlay Configurator
Measurement Time	500 s	Churn Generator
Transition Time	100 s	Churn Generator

4   Results

The performances of both the overlay protocols, Chord and Kademlia were evaluated against different varying parameters viz. mobility speed of nodes, lifetime of overlay nodes, overlay maintenance interval and network size under low mobility and low churn conditions. The confidence intervals for the parameters - mobility speeds, node lifetimes and network size have been chosen as suited to low mobility and low churn scenarios, like airports, railway stations, conference halls, workplaces, university campuses, etc. The overlay maintenance intervals for Chord and Kademlia have been varied by increasing the interval values from the standard values defined for both the protocols in OverSim, in order to reduce the frequency of exchanging overlay maintenance messages. The results obtained are based on the performance metrics (mean delivery ratio, mean latency and overlay maintenance overhead) measured against the network size.

4.1   Varying Mobility Speed of Nodes

Both Chord and Kademlia gave satisfactory performance under low mobility speeds of 0.1 m/s, 0.5 m/s and 2 m/s. The mean lifetime and mean deadtime of overlay nodes were set to 1000 s to maintain a low churn rate. Around 20% drop in the performance of Chord was seen when the network size was increased from 10 nodes to 200 nodes at all the three node speeds. Figure 2a shows the effect of growing network size on mean delivery ratio and Fig. 2b shows the increase in mean latency with growing network size. This performance slowdown was due to the increased overhead of overlay maintenance messages exchanged between overlay nodes in the dynamically growing Chord ring.

Kademlia, on the other hand exhibited a better performance than Chord. The performance of Kademlia in terms of mean delivery ratio was almost 100% at all node speeds (Fig. 3a), but a marginal performance drop in terms of increase in mean latency (0.1 s) with growing network size was observed (Fig. 3b).

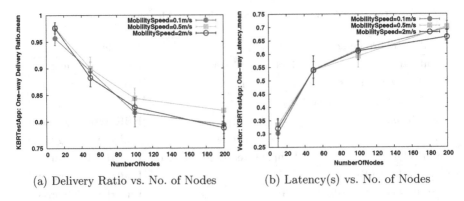

(a) Delivery Ratio vs. No. of Nodes   (b) Latency(s) vs. No. of Nodes

Fig. 2. Performance of Chord with varying node speeds (0.1m/s, 0.5m/s, 2m/s)

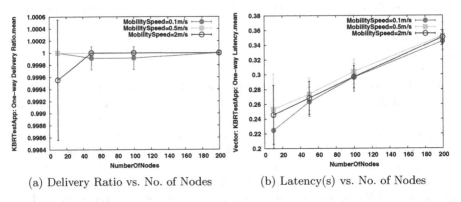

(a) Delivery Ratio vs. No. of Nodes   (b) Latency(s) vs. No. of Nodes

Fig. 3. Performance of Kademlia with varying node speeds (0.1m/s,0.5m/s,2m/s)

Hence, it is evident from the above results, that in a low mobility and low churn scenario, the mobility speed of overlay nodes does not have much influence on the performance of overlay application. However, large number of overlay nodes in the area slightly degrades the performance, because of increase in the overhead of exchanging overlay maintenance messages.

4.2 Varying Lifetime of Nodes

In order to observe the performance of overlay protocols over the dynamically changing topology of MANET for different lifetimes of overlay nodes, the mean lifetime was varied (500 s, 1000 s and 5000 s) to control the churn rate, but the mean deadtime was fixed at 500 s. Mobility speeds of overlay nodes were generated randomly by the Random Waypoint Mobility model in the interval (0.1 m/s, 2 m/s). The mean delivery ratio of Chord dropped significantly at mean lifetime of 500 s (from 94% to 58%) with growing network size, however on increasing the mean lifetime to 1000 s the performance improved, and eventually at mean lifetime of 5000 s approximately 100% delivery ratio was obtained for

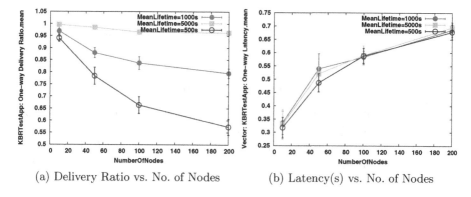

(a) Delivery Ratio vs. No. of Nodes   (b) Latency(s) vs. No. of Nodes

Fig. 4. Performance of Chord with varying mean lifetimes (500 s, 1000 s, 5000 s)

all network sizes, Fig. 4a. Identical performance curves in terms of mean latency were obtained for all three lifetimes, but a slight increase in latency was seen with larger lifetime values, Fig. 4b.

Kademlia, on the other hand gave better performance than Chord, as the delivery ratio was around 100% at all the network sizes for mean lifetimes of 1000 s and 5000 s, Fig. 5a. A marginal drop was observed at network sizes of 10 and 200, when the mean lifetime was 500 s. A small performance drop was also seen in terms of mean latency at lesser mean lifetime (500 s), Fig. 5b. But, the increase in latency with growing network size was less (0.1 s) in Kademlia than the increase shown in case of Chord (0.4 s).

Therefore, it can be said that for larger lifetimes of overlay nodes, both the overlay applications perform reasonably well. Longer lifetimes of nodes imply longer stay of the mobile nodes in the scenario, resulting in low churn rate. Maintaining low churn in the overlay stabilizes the topology maintained by the overlay applications. However, lesser lifetime of nodes increases the churn rate,

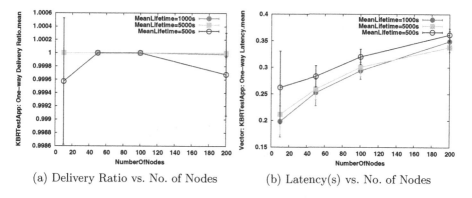

(a) Delivery Ratio vs. No. of Nodes   (b) Latency(s) vs. No. of Nodes

Fig. 5. Performance of Kademlia with varying mean lifetimes (500 s, 1000 s, 5000 s)

resulting in a dynamically changing topology, and therefore a performance drop. Also, the overall performance of Kademlia was better than Chord, thus establishing the suitability of Kademlia as an overlay protocol in such scenarios.

4.3 Varying Overlay Maintenance Interval

Here, we take different values of the overlay maintenance intervals to analyze the performance of the overlay applications in a low mobility and low churn scenario (default values of maintenance intervals were considered while varying mobility speed and mean lifetime of nodes). Both, the mean lifetime and the mean deadtime values were kept at 1000 s to maintain a low churn rate, and the mobility speeds of nodes were generated randomly in the interval (0.1 m/s, 2 m/s).

In case of Chord, the overlay maintenance interval was controlled by parameters *stabilizeDelay*, *fixfingersDelay* and *checkPredecessorDelay*, whose values were varied in proportion for different configurations of Chord. Following

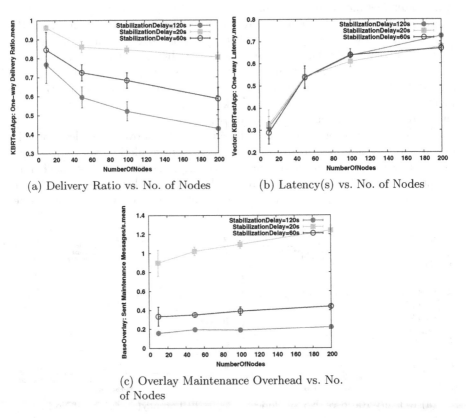

(a) Delivery Ratio vs. No. of Nodes   (b) Latency(s) vs. No. of Nodes

(c) Overlay Maintenance Overhead vs. No. of Nodes

Fig. 6. Performance of Chord with varying overlay maintenance intervals (20 s, 60 s, 120 s)

sets of values were taken - {*stabilizeDelay*=20 s, *fixfingersDelay*=120 s, *check-PredecessorDelay*=5 s}, {*stabilizeDelay*=60 s, *fixfingersDelay*=360 s, *checkPredecessorDelay*=15 s} and {*stabilizeDelay*=120 s, *fixfingersDelay*=720 s, *checkPredecessorDelay*=30 s}. The three configurations are shown in Fig. 6 in terms of stabilization delay.

The mean delivery ratio dropped with increasing network size in all the three configurations, but the performance of Chord was better at smaller intervals, Fig. 6a. The performance drop at 20 s stabilization delay was around 10% with growing network size, which increased to 20% at stabilization delay of 60 s and then to 35% at stabilization delay of 120 s. The mean latency increased with growing network size for all the three configurations, yielding similar performance curves with a slight increase in latency at network size of 200 for 120 s stabilization delay, Fig. 6b. Here, the performance drop can be explained in terms of overhead of maintenance messages being exchanged periodically between the overlay nodes. Frequent stabilization intervals resulted in increase of maintenance messages over a period of time, thus increasing the overhead, Fig. 6c.

However, as expected, Kademlia gave much better performance than Chord. For Kademlia, the overlay maintenance interval was controlled by parameters *minBucketRefreshInterval* and *minSiblingTableRefreshInterval*. Both the refresh interval values were kept equal in all the three configurations (500 s, 1000 s and 2000 s). The mean delivery ratio obtained by Kademlia was around 99% to 100% for all network sizes, Fig. 7a. The performance curves of mean latency were similar, and they indicated increase in latency with growing network size, Fig. 7b. The overhead of exchanging maintenance messages increased with growing network size, however increasing the frequency of stabilization did not have much effect on the performance of Kademlia, Fig. 7c.

Hence, it is evident that increasing the frequency of overlay maintenance procedures, increases the overhead but improves the delivery ratio of overlays. Chord performs reasonably well under low churn and low mobility conditions despite frequent maintenance procedures, but Kademlia gives better results in such scenarios, due to infrequent maintenance intervals required by its DHT.

4.4   Varying Network Size

All the performance results obtained above were recorded with different variations in mobility speed of nodes, lifetime of nodes and overlay maintenance interval against varying network sizes. The network size variations were taken along the X-axis in all the plots, with individual performance curves representing the variations in mobility speed, lifetime and overlay maintenance interval, Figs. 2, 3, 4, 5, 6 and 7. An overall performance degradation in terms of drop in delivery ratio, increase in latency and increase in maintenance overhead was observed for both Chord and Kademlia with large network sizes. However, Kademlia performed reasonably well as compared to Chord when network size was scaled up. These results were recorded for less dynamic MANETs, having low mobility speed, low churn rate and limited network size. Results obtained in [11, 12, 21] have already shown that, under highly dynamic scenarios the layered approach

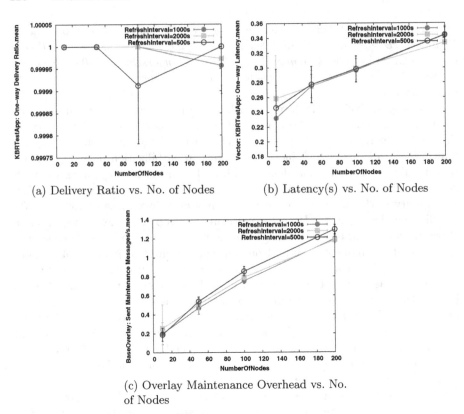

(a) Delivery Ratio vs. No. of Nodes   (b) Latency(s) vs. No. of Nodes

(c) Overlay Maintenance Overhead vs. No. of Nodes

Fig. 7. Performance of Kademlia with varying overlay maintenance intervals (500 s, 1000 s, 2000 s)

Table 2. Performance results of Chord over highly dynamic MANETs

Number of nodes	Mean one-way delivery ratio	Mean one-way latency
10	65%	0.325 s
50	21%	0.435 s
100	14%	0.465 s
200	6%	0.460 s

experiences a significant amount of performance degradation. In order to compare our results obtained for low dynamism in MANETs with that of moderately high dynamism in MANETs, we also simulated the behaviour of Chord under moderately high mobility and churn conditions. Same mobility and churn models were used for the simulation, i.e. Random Waypoint Mobility and Lifetime Churn models. The mobility speed of nodes was taken as 5 m/s for moderately high mobility, and both the mean lifetime and mean deadtime of nodes were kept as 100 s to maintain high churn dynamics. The results obtained (shown in

Table 2) establish that performance of Chord overlay was low under high mobility and high churn conditions, as evident from the drop in mean delivery ratio and increase in mean latency.

5 Conclusion and Future Work

In this paper, we have evaluated the performance of two different structured overlay applications over a less dynamic MANET using the layered approach. The results show that both Chord and Kademlia perform reasonably well, when deployed over MANETs with moderate number of mobile nodes, under low mobility and low churn conditions. However, with large network sizes a degradation in performance occurs due to increased overhead of exchanging overlay maintenance messages. Increasing the frequency of overlay maintenance procedures helps in keeping the DHT structure up-to-date, resulting in less amount of loss of data packets. However, exchanging maintenance messages at frequent intervals induces heavy traffic on the overlay network, thus increasing the overhead significantly. The layered approach for deploying structured peer-to-peer overlays is an elegant alternative for such scenarios, as it does not enforce any changes in the routing layer and MAC layer of participating nodes. This facilitates easy and rapid application development over MANETs. We have argued and established the utility of this approach for situations with less mobility and low churn rate.

Our proposal fits well in scenarios having low dynamism like, airports, railway stations, hotels, workplaces, university campuses, etc. However, these places may have separate zones like, waiting lounges, restaurants, cafetarias, etc., that can be interconnected via WiFi infrastructure, thus forming a hybrid and larger wireless network scenario. To merge these zones in a single DHT structure, a multi-tier architecture of the structured overlay application is required, on which we are working currently. We are also in process to develop adaptable maintenance methods of structured P2P overlays, that will take into account the underlying MANET characteristics, and will set the time periods of periodic maintenance messages accordingly.

References

1. Abid, S., Othman, M., Shah, N., Ali, M., Khan, A.: 3D-RP: a DHT-based routing protocol for MANETs. Comput. J. **58**(2), 258–279 (2015)
2. Abid, S.A., Othman, M., Shah, N.: A survey on DHT-based routing for large-scale mobile ad hoc networks. ACM Comput. Surv. **47**(2), 20:1–20:46 (2014)
3. Baumgart, I., Heep, B., Krause, S.: OverSim: a flexible overlay network simulation framework. In: Proceedings of 10th IEEE Global Internet Symposium (GI 2007) in Conjunction with IEEE INFOCOM 2007, Anchorage, AK, USA, pp. 79–84, May 2007
4. Benter, M., Divband, M., Kniesburges, S., Koutsopoulos, A., Graffi, K.: Ca-Re-Chord: a churn resistant self-stabilizing chord overlay network. In: 2013 Conference on Networked Systems (NetSys), pp. 27–34, March 2013

5. Binzenhöfer, A., Leibnitz, K.: Estimating churn in structured P2P networks. In: Mason, L., Drwiega, T., Yan, J. (eds.) ITC 2007. LNCS, vol. 4516, pp. 630–641. Springer, Heidelberg (2007). doi:10.1007/978-3-540-72990-7_56

6. Burresi, S., Canali, C., Renda, M., Santi, P.: Meshchord: A location-aware, cross-layer specialization of chord for wireless mesh networks (concise contribution). In: 6th Annual IEEE International Conference on Pervasive Computing and Communications, PerCom 2008, pp. 206–212, March 2008

7. Caesar, M., Castro, M., Nightingale, E.B., O'Shea, G., Rowstron, A.: Virtual ring routing: network routing inspired by DHTs. SIGCOMM Comput. Commun. Rev. 36(4), 351–362 (2006)

8. Caleffi, M., Paura, L.: M-DART: multi-path dynamic address routing. Wirel. Commun. Mob. Comput. 11(3), 392–409 (2011)

9. Castro, M., Kassler, A., Chiasserini, C.F., Casetti, C., Korpeoglu, I.: Peer-to-peer overlay in mobile ad-hoc networks. In: Shen, X., Yu, H., Buford, J., Akon, M. (eds.) Handbook of Peer-to-Peer Networking, pp. 1045–1080. Springer, Heidelberg (2010)

10. Castro, M., Villanueva, E., Ruiz, I., Sargento, S., Kassler, A.: Performance evaluation of structured P2P over wireless multi-hop networks. In: 2nd International Conference on Sensor Technologies and Applications, SENSORCOMM 2008, pp. 796–801, August 2008

11. Chowdhury, F., Kolberg, M.: Performance evaluation of structured peer-to-peer overlays for use on mobile networks. In: 2013 6th International Conference on Developments in eSystems Engineering (DeSE), pp. 57–62, December 2013

12. Cramer, C., Fuhrmann, T.: Performance evaluation of Chord in mobile ad hoc networks. In: Proceedings of 1st International Workshop on Decentralized Resource Sharing in Mobile Computing and Networking, MobiShare 2006, pp. 48–53. ACM, New York (2006)

13. Delmastro, F.: From pastry to crossroad: cross-layer ring overlay for ad hoc networks. In: 3rd IEEE International Conference on Pervasive Computing and Communications Workshops, PerCom 2005 Workshops, pp. 60–64, March 2005

14. Fuhrmann, T., Di, P., Kutzner, K., Cramer, C.: Pushing chord into the underlay: scalable routing for hybrid MANETs. Interner Bericht 2006-2012, Fakultät für Informatik, Universität Karlsruhe, 21 June 2006

15. Hu, Y.C., Das, S.M., Pucha, H.: Exploiting the synergy between peer-to-peer and mobile ad hoc networks. In: Proceedings of 9th Conference on Hot Topics in Operating Systems, HOTOS 2003, vol. 9, p. 7. USENIX Association, Berkeley (2003)

16. INETMANET 2.0. https://github.com/aarizaq/inetmanet-2.0

17. Jacquet, P., Muhlethaler, P., Clausen, T., Laouiti, A., Qayyum, A., Viennot, L.: Optimized link state routing protocol for ad hoc networks. In: Proceedings of Multi Topic Conference, IEEE INMIC 2001, Technology for the 21st Century, pp. 62–68. IEEE International (2001)

18. Johnson, D.B., Maltz, D.A., Broch, J.: DSR: the dynamic source routing protocol for multihop wireless ad hoc networks. In: Perkins, C.E. (ed.) Ad Hoc Networking, pp. 139–172. Addison-Wesley Longman Publishing Co., Inc., Boston (2001)

19. Krishnamurthy, S., El-Ansary, S., Aurell, E., Haridi, S.: A statistical theory of Chord under churn. In: Castro, M., Renesse, R. (eds.) IPTPS 2005. LNCS, vol. 3640, pp. 93–103. Springer, Heidelberg (2005). doi:10.1007/11558989_9

20. Maymounkov, P., Mazières, D.: Kademlia: a peer-to-peer information system based on the XOR metric. In: Druschel, P., Kaashoek, F., Rowstron, A. (eds.) IPTPS 2002. LNCS, vol. 2429, pp. 53–65. Springer, Heidelberg (2002). doi:10.1007/3-540-45748-8_5

21. Oliveira, L., Siqueira, I., Macedo, D., Loureiro, A., Wong, H., Nogueira, J.: Evaluation of peer-to-peer network content discovery techniques over mobile ad hoc networks. In: 6th IEEE International Symposium on World of Wireless Mobile and Multimedia Networks, WoWMoM 2005, pp. 51–56, June 2005

22. OMNeT++ 4.6. https://omnetpp.org/omnetpp

23. Perkins, C.E., Royer, E.M.: Ad-hoc on-demand distance vector routing. In: Proceedings of 2nd IEEE Workshop on Mobile Computer Systems and Applications, WMCSA 1999, p. 90. IEEE Computer Society, Washington, DC (1999)

24. Pouwelse, J., Garbacki, P., Epema, D., Sips, H.: The Bittorrent P2P file-sharing system: measurements and analysis. In: Castro, M., Renesse, R. (eds.) IPTPS 2005. LNCS, vol. 3640, pp. 205–216. Springer, Heidelberg (2005). doi:10.1007/11558989_19

25. Pucha, H., Das, S., Hu, Y.: Ekta: an efficient DHT substrate for distributed applications in mobile ad hoc networks. In: 6th IEEE Workshop on Mobile Computing Systems and Applications, WMCSA 2004, pp. 163–173, December 2004

26. Ripeanu, M.: Peer-to-peer architecture case study: Gnutella network. In: Proceedings of 1st International Conference on Peer-to-Peer Computing, pp. 99–100, August 2001

27. Rowstron, A., Druschel, P.: Pastry: scalable, decentralized object location, and routing for large-scale peer-to-peer systems. In: Guerraoui, R. (ed.) Middleware 2001. LNCS, vol. 2218, pp. 329–350. Springer, Heidelberg (2001). doi:10.1007/3-540-45518-3_18

28. Seddiki, M., Benchaïba, M.: An adaptive P2P overlay for MANETs. In: Proceedings of 2015 International Conference on Distributed Computing and Networking, ICDCN 2015, pp. 24:1–24:8. ACM, New York (2015)

29. Sözer, H.: A peer-to-peer file sharing system for wireless ad-hoc networks, August 2004

30. Stoica, I., Morris, R., Karger, D., Kaashoek, M.F., Balakrishnan, H.: Chord: a scalable peer-to-peer lookup service for internet applications. In: Proceedings of 2001 Conference on Applications, Technologies, Architectures, and Protocols for Computer Communications, SIGCOMM 2001, pp. 149–160. ACM, New York (2001)

31. Stutzbach, D., Rejaie, R.: Towards a better understanding of churn in peer-to-peer networks. Technical report, Univ. of Oregon (2004)

32. Trifa, Z., Khemakhem, M.: Effects of churn on structured P2P overlay networks. In: Proceedings of International Conference on Automation, Control, Engineering and Computer Science, ACECS 2014, pp. 164–170 (2014)

33. Zahn, T., Schiller, J.: MADPastry: a DHT substrate for practicably sized MANETs. In: Proceedings of 5th Workshop on Applications and Services in Wireless Networks (ASWN 2005), Paris, France, June 2005

Author Index

Printed in the United States
By Bookmasters